Millets - Rediscover Ancient Grains

Edited by Latika Yadav and Upasana

Published in London, United Kingdom

Millets - Rediscover Ancient Grains
http://dx.doi.org/10.5772/intechopen.106125
Edited by Latika Yadav and Upasana

Contributors
Brijesh Kumar, Magan Singh, Avaneesh Kumar, Deepak Kumar, Rakesh Kumar, Ankit Kumar Singh, Sandeep Kumar, K.B. Palanna, T. Tharana Poonacha, C.H. Sai Bhavana, Farooqkhan, G.V. Ramesh, Netravati Gavayi, Prasanna S. Koti, H. Rajashekara, G. Rajesh, I.K. Das, Smita Rana, Narendra Singh Bhandari, Alexander Bombom, Faitwa Walugembe, Tadeo Kaweesi, Sandiso Bhebhe, Mcebisi Maphosa, Sreenivasareddy Kadapa, Alekhya Gunturi, G. Raja Reddy, Srikanth Reddy Kalwala, Uday Bhaskar Mogallapu, Latika Yadav, R.K. Anushree, Meenal, Shailja Durgapal

First published in London, United Kingdom, 2024 by IntechOpen
IntechOpen is the global imprint of INTECHOPEN LIMITED, registered in England and Wales, registration number: 11086078, 5 Princes Gate Court, London, SW7 2QJ, United Kingdom

British Library Cataloguing-in-Publication Data
A catalogue record for this book is available from the British Library

Additional hard and PDF copies can be obtained from orders@intechopen.com

Millets - Rediscover Ancient Grains
Edited by Latika Yadav and Upasana
p. cm.
Print ISBN 978-1-83768-700-8
Online ISBN 978-1-83768-701-5
eBook (PDF) ISBN 978-1-83768-702-2

For EU product safety concerns:
IN TECH d.o.o., Prolaz Marije Krucifikse Kozulić 3, 51000 Rijeka, Croatia,
info@intechopen.com or visit our website at intechopen.com.

Meet the editors

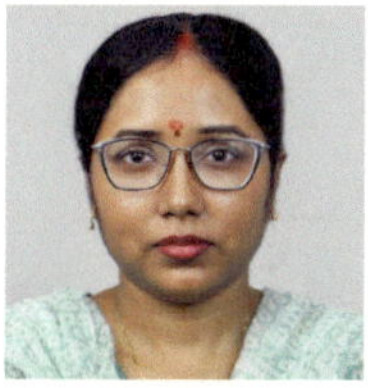

Dr. Latika Yadav is an assistant professor at Vijay Singh Pathik Government PG Degree College, Kairana, Shamli, Uttar Pradesh, India, with a strong background in academics, teaching, and research. She completed doctoral research funded by the National Fellowship-OBC, University Grant Commission (UGC), India. She has worked as a senior research fellow on a Department of Science and Technology (DST) project on the development of designer health foods and their promotion among rural women for the prevention and management of non-communicable diseases. She is an expert in nutrition science, food science, food product development, nutritional biochemistry, therapeutic diet, and public health nutrition. She has more than 7 years of teaching and more than 10 years of research experience in food science and nutrition. She has published numerous research and review papers in international and national journals. She has authored and co-authored several book chapters and edited and co-edited books.

Dr. Upasana is a highly qualified educator and researcher, currently serving as an assistant professor at Mahila College Dalmianagar, Rohtas, India. A gold medalist in post-graduation, she holds a Ph.D. from Banaras Hindu University, Varanasi, India, and is an expert in therapeutic nutrition, nutritional biochemistry, food science, microbiology, and public health nutrition. Her research work has been published in several prestigious academic journals. She has authored and co-authored several book chapters as well as co-edited one book. She serves as an editorial board member of an online journal and a reviewer for several peer-reviewed journals. Above all, Dr. Upasana is dedicated to providing her students with a comprehensive and engaging learning experience.

Contents

Preface

Millets are ancient grains that have been consumed for thousands of years around the world, especially in Africa and Asia. These grains are highly nutritious and contain a variety of vitamins, minerals, and antioxidants that are essential for human health. In addition to being a rich source of protein and fiber, millets are also gluten-free, making them an ideal food for people with gluten intolerance or celiac disease.

However, despite their numerous health benefits, these grains have been largely forgotten in modern times due to the dominance of other crops such as wheat, rice, and corn. As a result, the cultivation of millets has declined significantly in many regions.

Recently, there has been a renewed interest in millets as a sustainable and healthy alternative to other cereals. Millets are highly adaptable to different growing conditions, require low levels of water and fertilizer, and are resistant to pests and diseases. They offer farmers a viable option for sustainable agricultural practices.

Furthermore, the consumption of millets is not only beneficial for health but also for the environment. By consuming millets, individuals can reduce their carbon footprint and contribute to a more sustainable food system.

In conclusion, millets are ancient grains that have been rediscovered as a healthy and sustainable food source. By promoting their consumption and cultivation, we can improve our health and the health of our planet.

Latika Yadav
Assistant Professor,
Department of Home Science,
Vijay Singh Pathik Government PG Degree College,
Kairana, Shamli, Uttar Pradesh, India

Upasana
Assistant Professor,
Department of Home Science,
Mahila College Dalmianagar,
Rohtas, India

Chapter 1

Millets: Traditional "Poor Man's" Crop or Future Smart Nutri-Cereals?

Alexander Bombom, Tadeo Kaweesi, Faitwa Walugembe, Sandiso Bhebhe and Mcebisi Maphosa

Abstract

Millets represent a diverse group of cereal crops of significance to sub-Saharan Africa and globally. However, they remain a set of crops with limited attention and priority paid to them with paucity of information on their genetic diversity and sustainable use. Existing knowledge on millets with respect to cultivation, health, and nutritional benefits, and contribution to sustainable environmental management, and use is mainly attributed to traditional indigenous knowledge held by rural folks in different regions of the continent. The emergence of other cereal staples, however, led to millets losing their place as an important crop limiting their use to a "famine" crop with production occurring on smallholdings among the marginalized poor. This threatens interest, patronage, conservation and use to sustainably and fully exploit the potential of millets for the benefit of society. Intertwined with tradition and culture, millets in sub-Saharan Africa and elsewhere nonetheless hold great promise to contribute to food security, revitalize and diversify diets, improve farmer livelihoods, resilience, and adaptation to climate change. This chapter discusses the importance of millets, challenges to production, contribution to nutrition and health, traditional knowledge and products, novel and non-traditional products, contribution to resilience and climate change, and diversity of available genetic resources.

Keywords: millets, small grains, cereals, nutrition, health, climate resilience, value-addition, utilization, finger millet, pearl millet, sub-Saharan Africa

1. Introduction

Millets are an important set of cereal crops belonging to the family Poaceae. They have a wide distribution and are associated with cultures and food festivals among indigenous peoples of Africa and Asia. Of significance is the association of millets with food and nutrition security in marginalized arid and semi-arid regions of Sub-Saharan Africa (SSA) and elsewhere. A diversity of millets are cultivated in SSA, Asia and the Americas including but not limited to pearl millet [*Pennisetum glaucum* (L.) R. Br.], foxtail millet (*Setaria italica* (L.) P. Beauvois), little millet (*Panicum sumatrense* Roth.ex. Roem. & Schult.), Japanese barnyard millet [*Echinochloa esculneta* (A. Braun) H. Scholz], kodo millet [*Paspalum scrobiculatum* L.], proso millet [*Panicum miliaceum* L.], finger millet [*Eleusine coracana* (L.) Gaertn], Indian Barnyard millet [*Echinochloa frumetacea* Link], fonio or ach (*Digitaria exilis* (Kippist) Stapf), black fonio (*Digitaria iburua* Stapf), guinea millet (Brachiaria deflexa), and tef [*Eragrostis tef* (Zucc.) Trotter] [1]. Within the SSA context, four millets, that is, pearl millet, finger millet, fonio and teff are prevalent and of significance depending on geographical location – that is, east, central, southern and west Africa. In these regions of sub-Saharan Africa, millets contribute to the dietary energy and protein needs for at least 130 million people with significant impact on their food and nutrition security [1]. Of the four millets cultivated in Africa, pearl millet and finger millet stand out and have a wider distribution and global importance [2].

Pearl millet ranks sixth globally and is cultivated in areas with limited rainfall mainly in arid and semi-arid areas of SSA. Finger millet is majorly grown in more humid areas of SSA with its center of origin assumed to be in East Africa and more specifically, Uganda [3]. Traditionally, millets are cultivated by vulnerable groups including women and youths and are used as food security crops during times of scarcity. Furthermore, millets receive limited attention and investment for research and development, innovation and value addition, policy and political support. Oftentimes, referred to as a "poor man's crop", the importance and contribution of millets to the population in SSA and globally remains unnoticed. This is despite the innate, unexploited, biochemical, structural and nutritional characteristics potential of millets in food, feed, energy and industrial sectors. To fully harness and exploit millets for improved socio-economic and environmental benefit, increased effort towards development to address specialty market needs is imperative to drive demand and productivity. To achieve this within the context of climate change and variable weather patterns, self-reliance in food production in SSA will rely on low input agriculture in poor production environments.

2. Millets production trends in SSA

Women and youth comprise the primary producers of millets in SSA and together with other cereals provide the primary source of dietary energy for more than 962 million people. Nonetheless, production remains low, approximately 2.5 times below that of the developing world average characterized by subsistence level agriculture on small land holdings measuring 0.3–5.0 ha. A myriad of factors may contribute to this observation including extreme weather events (floods and drought), poor agronomic practices, lack of improved adapted genetic resources resistant to biotic and abiotic stresses, access to quality seed, traditional low yielding varieties, absence of appropriate smart and precision agriculture technologies for development and production of new adapted material, lack of well defined value chains for cultivated cereal crops, labour, bird damage and loss of essential crop biodiversity.

Production areas for millets span arid and semi-arid regions in east, south and west Africa. This has significant impact on rural socio-economic development, food security, feed security, energy security, health, and environmental sustainability for resource constrained persons in SSA. The major producing countries in sub-Saharan Africa include Nigeria, Niger, Burkina Faso, Senegal, Chad, Sudan, Mali, Uganda and Zimbabwe. In all of these countries, millet production is characterized by low input agricultural practices [4] in marginal lands with depleted soil fertility, high temperatures and drought [5]. Production volumes varied from 3038 to 3,886,079 tonnes per year from 2016 to 2020 with Niger producing the highest quantity of millet, followed by Nigeria, and Mali (**Figure 1**). Comparatively, production trends for millets in east

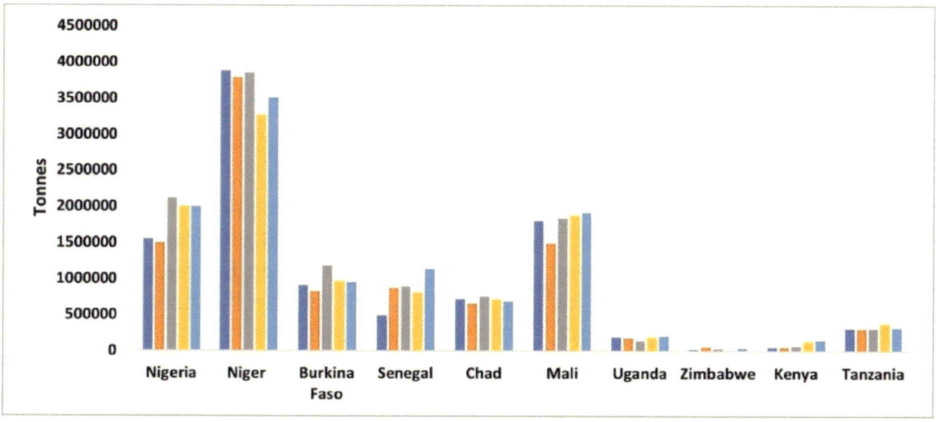

Figure 1.
Production quantity of millet from millet-producing countries in sub-Saharan Africa for the years 2016–2020 (FAOSTAT 2021).

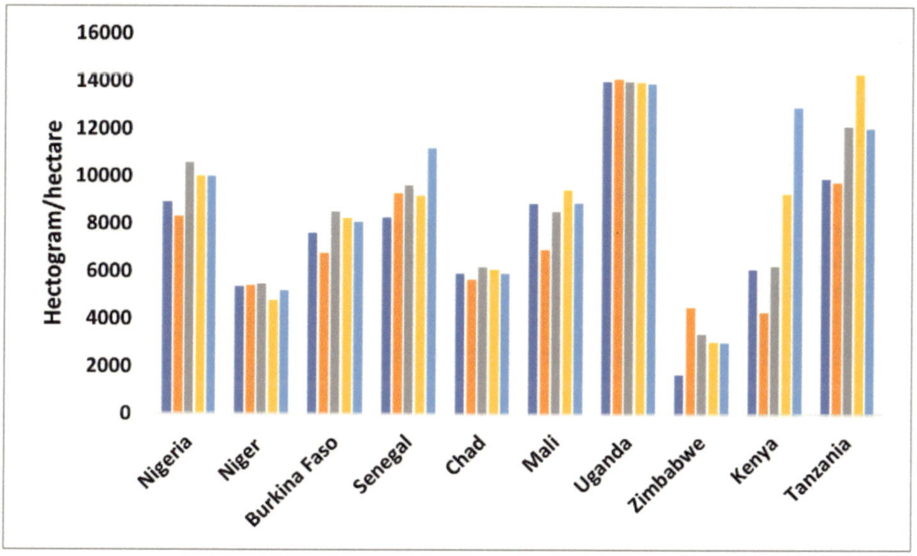

Figure 2.
Millet productivity in different millet-producing countries in sub-Saharan Africa for year, 2016–2020 (FAOSTAT 2021).

and southern Africa is low, with Tanzania ranked as the highest producer in the region followed by Uganda. Considering yield per unit area, East Africa has the highest millet productivity, with the highest yield (13,919–14,094 kg/ha) reported in Uganda from 2016 to 2020 (**Figure 2**). Other countries with high millet productivity in SSA include Tanzania, Senegal and Nigeria.

3. Why millets?

Despite the pivotal role and contribution to household food security, millets have over the years lost their place in today's cereal production systems and diets. This threatens interest, patronage, conservation and use to sustainably meet food security, farmer livelihoods and climate resilience for rural and urban communities alike. A number of factors could have contributed to the loss in popularity of millets relegating them to cultural food festivals and crops associated with famine hence, food for the poor. Furthermore, the decreasing trend in popularity and consumption of millets may be attributed to the availability of introduced cereals including maize, wheat, and rice, their associated public distribution system, easy methods of processing and preparation for consumption. The requirement for traditional knowledge and special skill mainly available among rural folks in preparing millet dishes and non-availability of ready-made value added products in the market further deter wider use of millets as food. Furthermore, the concept of the green revolution, promotion and adoption of alternate cereals such as hybrid maize varieties, access to credit, government support on maize prices and marketing subsidies also contributed to the decline in production of millets in the region [6]. Other challenges include the menace of quelea birds, drudgery and rising labour costs in production, emergence of fast foods and changing consumer preferences, widespread biotic stresses including striga infestation [7], diseases such as blast and downy mildew [7] and lack of remunerative market prices. Intertwined with tradition and culture, millets in sub-Saharan Africa and elsewhere nonetheless hold great promise to contribute to food security, revitalize and diversify diets, improve farmer livelihoods, adaptation to marginal soil conditions and climate resilience. The re-emergence in popularity of millets in SSA and elsewhere is due to the comparative advantages the crop offers in light of its potential contribution to climate change, urbanization and growing interest in nutrition and healthy foods. Millets adapt well to a wide range of climatic and marginal environments hence their compatibility with climate-smart agriculture interventions. Furthermore, the crops have a long post-harvest storage life and are seldom affected by storage pests [2].

3.1 Millets contribution to nutrition and health

Millets are a nutritionally valuable group of crops. Besides energy, millets provide a good source of quality protein and micronutrients. Pearl millet ranks highest in protein content (14.5%) among the millets, comparable to wheat (14.4%). Furthermore, pearl millet is associated with a more complete amino acid profile with valuable amino acids including methionine and tyrosine which are lacking in some millets [8, 9]. Both pearl millet and finger millet are rich in branched-chain amino acids (Leucine, isoleucine and valine) which have essential roles in protein synthesis, metabolic and regulatory processes [10, 11]. Finger millet possesses essential vitamins such as Vitamin A and B complex, water and fat-soluble vitamins [12]. Essential micronutrients present in pearl millet and finger millet include zinc, iron, calcium, sodium, potassium and magnesium [11]. Dietary consumption of millets facilitates vasodilation attributed to magnesium and potassium, which contributes to keeping the human body blood pressure at optimum [13, 14]. Consumption of finger millet has been reported to reduce plasma triglycerides, which helps reduce the risk of cardiovascular disease [15]. Furthermore, millets are rich in polyphenols and tannins, which contribute to reducing the risk for different kinds of cancer among populations consuming millets in their diets [16]. This observation is attributed to high radical scavenging or quenching activity in finger millet [17, 18]. Millets are a rich source of antioxidants such as curcumin, quercetin and ellagic acid which act as immune boosters, and important in the health and wellbeing of children and young adults [19].

As such, millets are of significance to integrated feeding and nutrition programs aimed at managing malnutrition and hidden hunger in SSA [20].

Millets have considerable health benefits given that they are reported to be gluten free [21]. As such, they are considered suitable for coeliacs. Coeliac disease is a syndrome characterized by damage to the mucosa of the small intestine caused by ingestion of gliadins and glutenins associated with wheat, rye and barley [21]. The prevalence of coeliac disease is estimated as high as 1:266 [21]. In the United States of America, persons with coeliac disease are estimated to be approximately 3 million. On the contrary, there are limited or no cases of coeliac disease in SSA. This may be because coeliac disease goes un-noticed and or cases are not reported or, the reliance of the rural population the majority of whom rely on millets. Furthermore, traditional knowledge across SSA has it that millets are natural nutraceuticals with meals often prepared and offered to persons with diabetes. Despite the association of millets with nutraceutical attributes even among local folks, there still remains a paucity of medical research of millets on human health with respect to the effects of phytochemicals and other components of the small grain cereals.

3.2 Traditional knowledge and products from millets

Across sub-Saharan Africa, traditional uses of millets are limited in scope mainly for food or local beverages and or brew. Part of the limitation in use is also attributed to the requirement of special skill in preparing and processing millets, which knowledge is not common place. This tends to limit the demand for millets, particularly in urban areas where consumers have access to alternate processed value-added foods. Promotion and food enhancement of millets can contribute to improve the nutritional status and livelihoods of poor marginalized communities. Traditional technologies such as fermentation, malting, soaking, milling/grinding, cooking, roasting and popping are common place in SSA [22]. Each of these processing technologies improve the nutritional value, functional qualities and sensory properties of millets. For example, soaking reduces anti-nutritional compounds such as trypsin inhibitors, phytate, tannins, phytic acids and flavonoids, thus increasing the bioavailability of minerals such as zinc and iron [23]. Germination and malting facilitate hydrolysis of proteins and starch, making amino acids and sugars readily available in processed products, improves digestibility and availability of vitamins and minerals [24, 25].

Preparation and processing of traditional millet products varies by region and cultures within SSA. The product range may include but not be limited to creamy millet drink (*Kanu-zaki* in Nigeria), fermented millet porridge or gruel (*Ogi* in Nigeria, *Ben-saalga*, *Ben-kida* in Burkina Faso, *Malwa* or *Ajon* in Uganda), unfermented millet porridge or gruel (*Bushera* in Uganda, *Uji* in Kenya or Tanzania, *Togwa* in Tanzania), and millet bread or dough (*Kalo* in Uganda, *Tehobal* and *gappal* in Burkina Faso) (**Table 1**). In Ethiopia, gluten-free leavened breads (injera) may be prepared from millets [21].

With improvements and access to technology, across regions in SSA, various other value-added millet products have been formulated using several components such as yoghurt, milk, ginger, chili, other cereal blends, baobab pulp or leaves, tamarind juice, fruits such as pineapple and spices. The combination improves both the taste/sensory attributes as well as the nutritive value of the formulated millet products. It is worth noting that some nutrients such as vitamins may be reduced or lost during the processing of millets into other ready-to-drink products [26]. The diversity of millet products especially in West Africa shows that the utilization

Region	Country	Product	Description	Recipe/composition
West Africa	Nigeria	*Masa*	Fried cake	Millet, yeast, yoghurt, sugar, salt, onion
		Kunu-zaki	Creamy millet drink	Millet, sugar, germinated rice, cloves, ginger, potato, milk
		Fura	Boiled millet balls	Millet, cloves, ginger, chili pepper
		Ogi	Fermented millet porridge	Millet, water, fermentation culture
	Burkina Faso	*Massa*	Millet pancake	Millet, rice, Maize Tô, oil, water
		Ben-saalga	Fermented millet gruel	Millet, ginger, black pepper, mint, water
		Ben-kida	Fermented millet gruel	Millet flour, millet granules, ginger, pepper, mint, water
		Millet gruel	Millet gruel for Northern region	Millet, sugar, curd, cooking oil, salt, baobab pulp, water
		Zom-kom	Crude millet beverage	Millet, sugar, ginger, mint, Tamarind leaves, tamarind fruits, pineapple powder, canafra, water
		Millet pase	Thick millet dough	Millet flour, water, tamarind juice
		Tounka	Bran millet beverage	Millet flour, millet bran, sugar, salt, solid potash, water
		Tehobal	Millet dough	Millet flour, curd cheese, sugar, water
		Gappal	Millet dough with milk	Millet flour, curdled milk or yoghurt, sugar, salt, water
		Millet couscous	Granular steamed food	Millet flour, water, salt, baobab leaf or okra powder
	Senegal	*Thiakry*	Millet couscous topped with mangoes, raisin, coconuts & nuts	Millet, salt, water, milk, yoghurt, vanilla, cinnamon, chopped pistachios
	Mali	*Toh (Oro Dja)*	Traditional food of the Dogon people	Husked millet, water, dried okra, dried fish, dried baobab leaves, fermented hibiscus seeds
	South Sudan	*Medeeda*	Millet porridge with yoghurt	Millet, water, peanut, sugar, lime juice and yoghurt
	Cameroon	*African Millet parties*	Millet cookies	Millet, raisin, water, onion, celery seeds, soy source
East Africa	Uganda	*Bushera*	Millet porridge	Millet flour, water, milk, sugar
		Kalo	Millet bread	Millet flour, cassava flour, water
		Malwa/ajon	Millet beer	Germinated millet, water, sorghum
	Kenya/ Tanzania	*Uji*	Millet porridge	Millet flour, water, milk, sugar
		Togwa	Millet gruel	Germinated finger millet, maize flour, water

Region	Country	Product	Description	Recipe/composition
Southern Africa	Zimbabwe	*Mahewu*	Millet malt drink	Millet malt, water
		Sadza rezviyo/ Isitshwala samabele/ Inyawuthi	African millet pap	Millet, water
		Masvusvu	Millet mash	Malted millet, water

Table 1.
Traditional millet-based products and their composition from different sub-Saharan countries.

and demand for millet can be improved in SSA, which will directly benefit farmers, processors and other value chain actors. Across different African countries, similar millet products are given different names. For example, millet beer is named differently in different areas such as *Malwa* or *Ajon* (Uganda), *Merissa* (Sudan), *Chikokivana* or *Ithothotho* (Zimbabwe), *Chagga* (Tanzania), *Tchouk* (Togo) and *Dolo-Djioula* (Burkina Faso). There may be variations in the preparation of millet beer in these regions, however, standardization of this process offers an opportunity for promoting millet beer in a wider area, increasing the production and utilization of millets in SSA.

3.3 Novel and non-traditional millet products

Over the years, advances in food processing technologies and cereal sciences in general have presented opportunity for the development of novel and non-traditional products from diverse cereals including maize, wheat, rye, rice and barley. These advancements can be of significant value in the development of millets in the quest to improve and support their re-adoption, production and commercialisation in the face of growing health concerns and climate change.

Tortillas, snack foods, noodles, starch breads, flour breads and additives, cakes, cookies, and pasta are examples of novel products that have been successfully produced from other cereals. One hundred percent millet bread and/or confectionery products remains the main challenge. However, this challenge could be ameliorated with additives and enzymes to improve product quality.

As opposed to traditional African opaque beer, malting and brewing with millets to produce lager and stout, often referred to as clear beer, has not been conducted on a large, commercial scale. This is because brewing processes with millets have not been extensively investigated and are only at an experimental and/or piloting stage.

However, significant traditional beverages from millets exits diverse in recipe by culture and region within SSA. More recently, the Coca cola company launched the Ades global brand of plant-based beverages. In Uganda, the Coca cola company added Ades Nutri-Bushera to its product portfolio, which offers an opportunity to boost millet production, productivity and income for populations living in marginal arid and semi-arid areas of SSA [27]. Through cultivar selection and crop improvement, millets can be used to produce non-food products including bioethanol and other bio-industrial products such as bioplastics, thus contributing to rural circular bioeconomy in marginal areas [21].

3.4 Contribution of millets to climate change adaptation and resilience

Climate change remains a significant challenge to agriculture in SSA presenting with unpredictable weather, limited and erratic rainfall. However, millets hold promise to secure food and nutrition security for the approximately 1.17 billion inhabitants in SSA. Governments across SSA have put in place agriculture sector, Development Strategy and Investment Plans that seek to increase household incomes and improve food security by enhancing agricultural production and productivity, achieve economic growth through industrialization, reduce greenhouse gas emissions and become middle-income countries by the year 2050. In line with the African Union (AU) Agenda 2063, millets have the potential to contribute to the aspirations – "A prosperous Africa, based on inclusive growth and sustainable development" and "An Africa whose development is people driven, relying on the potential offered by people, especially its women and youth and caring for children".

The importance of millets as a climate-smart cereal in contributing to climate change adaptation, mitigation and resilience cannot be underestimated for a number of reasons. Farmers in SSA are familiar with its cultivation and is easy to propagate. Millets are adaptable to diverse environments and tolerates heat and drought stress. With decreasing water supplies, increasing population growth and their requirement for limited inputs during growth, millets represent an important set of genetic resource for future human use. The annual growth cycle of millets (2.5–4 months) does not tie up land that can be used for other uses and, can be rotated or intercropped with other food crops ensuring food security, improved soil integrity and minimizes disease and pest epidemics. The root system of millets facilitates below-ground carbon sequestration providing rapid carbon sink and lowering greenhouse gas emissions.

4. Millets genetic resources and improvement

Millets biodiversity is an invaluable resource to African agricultural production systems to help address unpredictable climatic conditions, household nutrition, and socio-economic challenges in SSA. This can be achieved through crop diversification, conservation, and sustainable use integrated with indigenous knowledge. About 133,849 cultivated germplasm of small millets are conserved in gene banks worldwide the majority being conserved in Asia (64.4%), followed by Africa (13.8%), and Europe (13.5%) [28]. More than 22,211 pearl millet and 35,382 finger millet have been collected and consolidated in different gene banks with the largest collection at ICRISAT. Of these, a few have been utilized as breeding/research material [29]. Despite these numbers, conservation of millets in genebanks in SSA and globally is generally compared with other crop species. It is plausible this might be due to the low research and priority accorded to millets and associated land races leading to loss in biodiversity. In Uganda for instance, pearl millet biodiversity has been reduced or even lost with farmers citing damage from birds as the primary reason for abandoning the crop despite its importance. To prevent further loss of existing, available diversity, it is imperative that concerted efforts in supporting collection, conservation and use of available millets' diversity is carried out before we lose them forever.

The year 2023 has been declared the "International Year of Millets" by UN General Assembly in recognition of its importance as a climate smart food crop with to contribute to alleviation of malnutrition, hunger and poverty. The latest UN projection suggests that the world population could grow to around 9.7 billion in 2050, with

Sub-Saharan Africa contributing over 50% of this population growth [30]. The intergovernmental panel on climate change warned that the global food supply is already jeopardized [31]. Therefore, resilience of millets to climate and weather variations makes it a suitable future smart Nutri-cereal. To achieve this, efforts in crop improvement and processing technologies are needed to generate "smart super millets" with specialty traits targeting industrial applications to increase demand, productivity and subsequently popularity of millets. A number of methods and tools are available including conventional crop improvement and marker assisted breeding approaches. Target traits of interest for crop improvement will include diseases and pest, drought and heat stress, nutrition quality for macro- and micronutrients, yield and water use efficiency. In livestock communities, biomass would be an important trait for dual purpose millet varieties to provide food and feed security.

More recently, the speed breeding approach has been used as a strategy to reduce the time frame for variety development in crops such as wheat. Speed breeding can be integrated with tools including high throughput phenotyping, genotyping by sequencing (GBS), genome editing technologies such as CRISPR-Cas9, and biochemical analyses to develop climate smart millets. Progress by the International Crops Research Institute for the Semi-Arid Tropics (ICRISAT) in millets improvement, has focused on pearl millet and finger millet [28] with development of millet hybrids high in yield, with high absorbable nutrients. A clear example is the work done by HarvestPlus, a CGIAR challenge program leading global efforts in the development of micronutrient-rich crop varieties. The program started in 2003, focusing on the biofortification of millet, bean, cassava, rice, sweet potato, wheat and maize [32, 33], contributing to the alleviation of malnutrition due to iron and zinc deficiency affecting 60% and 30% of the world population respectively [34]. Evidence of the impact of biofortified pearl millet with absorbable iron concentration of 60 ug/g has been documented [23]. So far, a total of 11 high iron pearl millet varieties, 9 single cross and 2 open-pollinated varieties have been released [33], only one variety has been released in Africa (in Niger). Efforts should be made to evaluate all 11 varieties in different millet producing countries for a decentralized selection of the best millet varieties for nutrition programs to address different health problems caused by iron deficiency [35, 36].

Open source web-based platforms are available with high-computing capabilities and comprehensive breeding management and analysis software such as Usegalaxy (http://usegalaxy.org), Breedbase (http://breedbase.org), integrated breeding platform (http://www.integratedbreeding.net), and Excellence in breeding platform (http://excellenceinbreeding.org) to facilitate accelerated crop improvement in low and middle-income countries. These efforts are complemented by genotyping services at regional hubs such as SEQART Africa that offers sequencing services in support of national breeding programmes. The availability of such service facilitates molecular breeding such as diversity studies, genome-wide association studies for unraveling of the genetic architecture of key traits, genomic selection and gene discovery for fast tracked breeding and increased genetic gain in different crops.

5. Conclusions and future perspectives

The re-emergence of millets as an important set of crops not just for SSA but globally demonstrates their potential to serve as alternate and/or supplement cereals along side major staples – maize, rice, wheat, and barley. Nonetheless, significant efforts still remain to bring millets at par with other cereals given they remain virtually unresearched and their potential unexploited. From a crop improvement perspective, understanding genetic diversity within and between millets defines the first step in their development for effective and sustainable utilization. To guide the process of millets improvement, farmers with traditional indigenous knowledge, industry and other stakeholders in the value chain need to participate as partners providing a platform for co-creation and development of value-added products subsequently driving consumer demand, incomes and rural economic development. To enhance the rate of genetic gain and improvement in millets, adoption of tools including the use of artificial intelligence, high throughput phenomics and genomics are imperative to assure timeliness and efficiency in selection. It is worthwhile noting that, climate-ready millets in themselves are not sufficient to achieve increased productivity, food security, rural incomes, resilience and adaptation to climate change. Rather, identifying specialty attributes in improved millets genetic resources that meet niche market needs coupled with circular economy approaches will address limitations of value chains and create supplier linkages for millions of small-scale farmers and consumers across SSA. To compliment this, as with other crops, increase in productivity of millets can become a reality if stakeholders in SSA can adopt precision agriculture technologies including but not limited to the use of unmanned arial vehicles to monitor crop health, growth and physiology; integrating laser scare crows alongside other technologies for bird management, farm management software, farm mechanization and postharvest processing.

To attract the support of different authorities including government to promote research, development and commercialisation of millets across Africa, a look into millets beyond just contributing to resilience and household food security is imperative. Deliberate efforts need to demonstrate the contribution of millets to health and nutrition, rural employment, household incomes, rural and national economies, and sustainable environment management. Such efforts may include but not be limited to policies support on partnerships on co-development of high value millet products with large customer bases; integrating gender equality aspects in the promotion of millets taking into consideration the contribution of rural women in genetic resource management, household food security, and their role in subsistence agriculture.

Acknowledgements

This article has been produced in the framework of the Benefit-sharing Fund project "PR-316-Uganda" that has been funded by the Food and Agriculture Organization's (FAO) International Treaty on Plant Genetic Resources for Food and Agriculture (ITPGRFA): https://www.fao.org/plant-treaty/areas-of-work/benefit-sharing-fund/projects-funded/bsf-details/en/c/1198836/?iso3=UGA. The views expressed in this article are those of the author(s) and do not necessarily reflect the views or policies of FAO.

Conflict of interest

The authors declare no conflict of interest.

Notes/thanks/other declarations

None.

Author details

Alexander Bombom[1,2]*, Tadeo Kaweesi[1], Faitwa Walugembe[1], Sandiso Bhebhe[3] and Mcebisi Maphosa[3]

1 Biosciences Research Programme, National Agricultural Research Organisation-National Livestock Resources Research Institute (NaLIRRI), Wakiso, Uganda

2 Bomvitae Agro Industries Limited (BAIL), Kampala, Uganda

3 Department of Crop and Soil Sciences, Lupane State University, Lupane, Zimbabwe

*Address all correspondence to: xander.bombom@gmail.com

References

[1] Obilana AB. Overview: Importance of Millets in Africa. Nairobi, Kenya: International Crops Research Institute for the Semi-Arid Tropics (ICRISAT); 2003

[2] Garí JA. Review of the African millet diversity. 2002. Available from: https://www.semanticscholar.org/paper/Review-of-the-African-millet-diversity-Gar%C3%AD/d0353ee29059c467cfd2e233d16d73b4cf2c2e4b. [Accessed: December 26, 2022]

[3] Serba DD, Perumal R, Tesso TT, Min D. Status of global pearl millet breeding programs and the way forward. Crop Science. 2017;**57**:2891-2905

[4] Mukarumbwa P, Mushunje A. Potential of sorghum and finger millet to enhance household food security in Zimbabwe's semi-arid regions: A review. African Association of Agricultural Economists (AAAE). 2010;**96430**:23

[5] Satyavathi CT, Ambawat S, Khandelwal V, Srivastava RK. Pearl millet: A climate-resilient Nutricereal for mitigating hidden hunger and provide nutritional security. Front Plant Sci. Frontiers Media S.A. 2021;**12**:1828

[6] Kanfany G, Diack O, Kane NA, Gangashetty PI, Sy O, Fofana A, et al. Implications of farmer perceived production constraints and varietal preferences to pearl millet breeding in Senegal. African Journals Online (AJOL). 2020;**28**:411-420

[7] Rouamba A, Shimelis H, Drabo I, Laing M, Gangashetty P, Mathew I, et al. Constraints to pearl millet (Pennisetum glaucum) production and farmers' approaches to Striga hermonthica Management in Burkina Faso. Sustainability. 2021;**13**:8460

[8] Ravindran G. Studies on millets: Proximate composition, mineral composition, and phytate and oxalate contents. Food Chemistry. 1991;**39**:99-107

[9] Ramashia SE, Anyasi TA, Gwata ET, Meddows-Taylor S, Jideani AIO. Processing, nutritional composition and health benefits of finger millet in sub-saharan Africa. Food Science and Technology. Sociedade Brasileira de Ciência e Tecnologia de Alimentos. 2019;**39**:253-266

[10] Kumar A, Metwal M, Kaur S, Gupta AK, Puranik S, Singh S, et al. Nutraceutical value of finger millet [Eleusine coracana (L.) Gaertn.], and their improvement using omics approaches. Front Plant Science. 2016;**7**:934

[11] Chandra D, Chandra S, Pallavi SAK. Review of finger millet (Eleusine coracana (L.) Gaertn): A power house of health benefiting nutrients. Food Science and Human Wellness. 2016;**5**:149-155

[12] DeviPB,VijayabharathiR,SathyabamaS, Malleshi NG, Priyadarisini VB. Health benefits of finger millet (Eleusine coracana L.) polyphenols and dietary fiber: A review. J Food Sci Technol. 2014;**51**:1021-1040

[13] Haenni A, Johansson K, Lind L, Lithell H. Magnesium infusion improves endothelium-dependent vasodilation in the human forearm. American Journal of Hypertension. 2002;**15**:10-15

[14] Park SY, Jeong EW, Yang YS, Kim HJ, Go GW, Lee HG. Finger millet ethanol extracts prevent hypertension by inhibiting the angiotensin-converting

enzyme level and enhancing the antioxidant capacity in spontaneously hypertensive rats. Antioxidants. 2021;**10**:1766

[15] Lee SH, Chung IM, Cha YS, Park Y. Millet consumption decreased serum concentration of triglyceride and C-reactive protein but not oxidative status in hyperlipidemic rats. Nutrition Research. 2010;**30**:290-296

[16] Grimmer HR, Parbhoo V, McGrath RM. Antimutagenicity of polyphenol-rich fractions from sorghum bicolor grain. Journal of the Science of Food and Agriculture. 1992;**59**:251-256

[17] Dykes L, Rooney LW. Sorghum and millet phenols and antioxidants. Journal of Cereal Science. 2006;**44**:236-251

[18] Chandrasekara A, Shahidi F. Bioaccessibility and antioxidant potential of millet grain phenolics as affected by simulated in vitro digestion and microbial fermentation. Journal of Functional Foods. 2012;**4**:226-237

[19] Dhaka A, Singh RK, Muthamilarasan M, Prasad M. Genetics and genomics interventions for promoting millets as functional foods. Current Genomics. 2021;**22**:154

[20] Maharajan T, Antony Ceasar S, Ajeesh Krishna TP, Ignacimuthu S. Finger millet [Eleusine coracana (L.) Gaertn]: An orphan crop with a potential to alleviate the calcium deficiency in the semi-arid tropics of Asia and Africa. Front Sustain Food Syst. 2021;**5**:258

[21] Taylor JRN, Schober TJ, Bean SR. Novel food and non-food uses for sorghum and millets. Journal of Cereal Science. 2006;**44**:252-271

[22] Saleh ASM, Zhang Q, Chen J, Shen Q. Millet grains: Nutritional quality, processing, and potential health benefits. Comprehensive Reviews in Food Science and Food Safety. 2013;**12**:281-295

[23] Tako E, Reed SM, Budiman J, Hart JJ, Glahn RP. Higher iron pearl millet (Pennisetum glaucum L.) provides more absorbable iron that is limited by increased polyphenolic content. Nutrition Journal BioMed. 2015;**14**:11

[24] Krishnan R, Dharmaraj U, Malleshi NG. Influence of decortication, popping and malting on bioaccessibility of calcium, iron and zinc in finger millet. LWT - Food Science and Technology. 2012;**48**:169-174

[25] Mbithi-Mwikya S, van Camp J, Yiru Y, Huyghebaert A. Nutrient and Antinutrient changes in finger millet (Eleusine coracan) during sprouting. LWT - Food Science and Technology. 2000;**33**:9-14

[26] Durojaiye AFA, Falade KO, Akingbala JO. Chemical composition and storage properties of fura from pearl millet (*Pennisetum americanum*). Journal of Food Processing & Preservation. 2010;**34**:820-830

[27] Coca-Cola adds Ades Nutri-Bushera to its product portfolio [Internet]. Available from: https://www.independent.co.ug/coca-cola-adds-ades-nutri-bushera-to-its-product-portfolio/ [Accessed: November 29, 2022]

[28] Vetriventhan M, Azevedo VCR, Upadhyaya HD, Nirmalakumari A, Kane-Potaka J, Anitha S, et al. Genetic and genomic resources, and breeding for accelerating improvement of small millets: Current status and future interventions. Nucleus (India). 2020;**63**:217-239

[29] Moreta DE, Mathur PN, Zonneveld M van, Amaya K, Arango J, Selvaraj MG, et al. Current issues in cereal crop

biodiversity. Advances in Biochemical Engineering/Biotechnology. 2013; **147**:1-35

[30] World population to reach 8 billion this year, as growth rate slows. UN News [Internet]. Available from: https://news.un.org/en/story/2022/07/1122272 [Accessed: November 30, 2022]

[31] Climate change a threat to security, food and humankind - IPCC report | Climate crisis. The Guardian [Internet]. Available from: https://www.theguardian.com/environment/2014/mar/31/climate-change-threat-food-security-humankind [Accessed: November 30, 2022]

[32] Bouis HE, Saltzman A. Improving nutrition through biofortification: A review of evidence from harvest plus, 2003 through 2016. Global Food Security. 2017;**12**:49-58

[33] Bouis H. Reducing mineral and vitamin deficiencies through biofortification: Progress under harvestplus. World Review of Nutrition and Dietetics. 2018;**118**:112-122

[34] White PJ, Broadley MR. Biofortification of crops with seven mineral elements often lacking in human diets--iron, zinc, copper, calcium, magnesium, selenium and iodine. New Phytologist. 2009;**182**:49-84

[35] James AH. Iron deficiency anemia in pregnancy. Obstetrics and Gynecology. NLM (Medline). 2021;**138**:663-674

[36] Bongomin F, Olum R, Kyazze AP, Ninsiima S, Nattabi G, Nakyagaba L, et al. Anemia in Ugandan pregnant women: A cross-sectional, systematic review and meta-analysis study. Tropical Medicine and Hygiene. 2021;**49**:1-13

Chapter 2

Nutritional Properties, Nutraceutical Potential of Different Millets, and Their Value-Added Food Products

Smita Rana and Narendra Singh Bhandari

Abstract

Millets are one of the oldest food grains known to mankind. They are considered underutilized crops and can sustain in harsh environments with limited water resources where other crops grow or yield poorly. Millets are very adaptable: they thrive in dry regions, on clay soils, in wet lowlands, or alluvial lands. Their root systems are powerful, able to descend very quickly to a great depth of soil to extract water and minerals, and thus have high adaptability to climate change, especially drought. Millets have a good nutritional profile along with good nutraceutical potential and thus can be considered a great crop for combating food nutritional security globally.

Keywords: nutrition, medicinal properties, millets, value-added products, food products

1. Introduction

Millets are considered to be one of the oldest and most important cereal grains known to mankind [1–3]. Millets are ranked as the 6th cereal crop in terms of the world's agricultural production and are the staple diet for nearly 1/3rd of the world's population [1, 2]. Millets constitute the staple diet in many parts of Asia and Africa since ancient times [4].

Millets are usually small-seeded annual cereal grains that are part of the Poaceae family, are cultivated throughout tropical and subtropical region [5]. Millets can be grown even in areas with limited natural resources because they need very little water for production and can be grown without irrigation or in areas with very little rainfall (200–500 mm) i.e., can withstand drought conditions, higher heat regimes, less fertile soil, pest resistance, and short growing season usually 45–60 days [2, 3, 5]. Millets are classified as C4 cereals. C4 cereals are more environmentally friendly because they use water more efficiently, convert more carbon dioxide from the atmosphere to oxygen, and require less energy [6]. Millets are broadly categorized into two major groups: (1) major millets; pearl millet or bajra (*Pennisetum glaucum*), sorghum or jowar (*Sorghum bicolor*) and (2) minor millet; finger Millet or ragi, mandua (*Eleusine coracana*), barnyard millet or jhangora, sanwa (*Echinochloa frumentacea*), little millet or kutki, shavan (*Panicum sumatrense*), foxtail millet or kangni, kakum (*Setaria italica*), proso millet or barri (*Panicum miliaceum*), and kodo millet or koden (*Paspalum scrobiculatum*) are the millet crops largely cultivated in Asian and African countries [7]. Globally, India is the largest producer of millet with 35.625% of the world's total production. Among the top ten millet-producing countries of the world, India ranked first with 1.085 MT followed by Nigeria (0.45 MT), Niger (0.28 MT), China (0.21 MT), Mali (0.127 MT), Sudan (0.108 MT), Burkina Faso (0.097 MT), Senegal (0.062 MT), and Ethiopia (0.058 MT) (**Figure 1**). The majority of millets, 95.9%, are produced in developing countries, mostly in Africa and Asia (**Figure 2**) [8]. In India, millets are grown in the states of Karnataka, Tamil Nadu, Andhra Pradesh, Kerala, Maharashtra, Telangana, Uttarakhand, Jharkhand, Chhattisgarh, Orissa, and Madhya Pradesh [9, 10].

Millets were discovered to have a high nutritional content that is equivalent to that of popular grains like wheat and rice [11]. They are considered as high-energy yielding nourishing foods which help in addressing malnutrition [5]. **Table 1** summarizes the average nutritious makeup of several millet grains and other significant cereal grains. Along with amazing nutritional values millets also have health benefits. A number of health benefits are associated with the consumption of millets, largely due to the bioactive phytochemicals found in these cereals, such as lignans, flavonoids, phenolics, beta-glucan, sterols, inulin, pigments, dietary fiber, and phytate [5]. They aid in the management of health issues such as diabetes mellitus, hyperlipidemia, cancer, and cardiovascular illnesses, as well as decreasing blood pressure, risk of heart disease, cholesterol, and the rate of fat absorption, and delaying stomach emptying [2, 11]. Phenolics rich millets are of great significance in health, aging, and metabolic syndrome. The presence of phytate in the millets is allied with anticancer and cholesterol lowering property. Millets with fiber-rich content are essential for preventing the gall stone formation. Millets are most commonly thought of as a good source of proteins which play a crucial role in the suppression of malnutrition [5]. Additionally, millets are also used as an important component of many traditional diets, and are

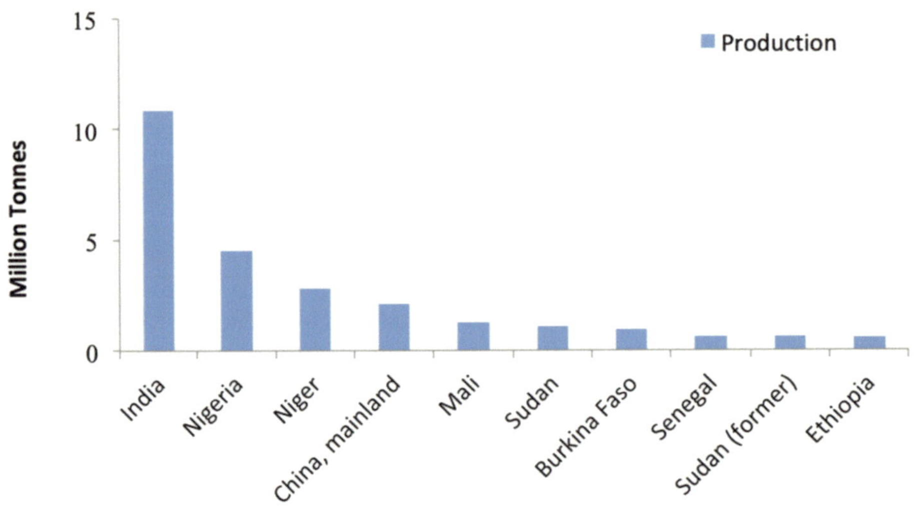

Figure 1.
Top 10 producers of millet in the world.

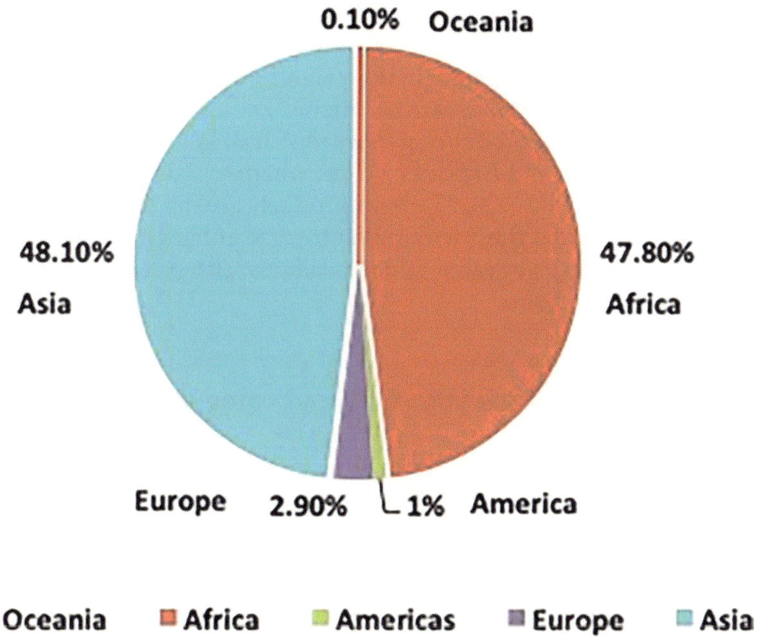

Figure 2.
Production share of millets by different region.

Grains	Protein	Carbohydrate	Fat	Fiber	Ash	Moisture	Calorific value (kcal)
Barnyard millet	10.76–13	55.7–74	3.5–4.8	3.9–13.6	3.3–4.6	7.78–11.24	300–310
Finger millet	7.3–10	71.52–83.3,	1.30–1.8	3.4–4.2	2.63–2.8	7.68–13.1	328–334
Pearl millet	10.6–11.8	59.8–75.6	4.8–5.7	1.3–2.3	1.64–2.2	12.4	363–412
Foxtail millet	11.34–12.3	60.2–75.2	3.33–4.3	4.1–8.7	3.37	7.69–11.2	330–352
Proso millet	11.74–13	67.09–82	1.1–4.9	2.2–8.47	2.73–4	11.9	330–352
Kodo millet	8.3–10.2	63.82–73.5	1.4–3.9	5.2–9.5	2.83–3.6	8.06–10.83	309–349.5
Little millet	7.7–10.7	66.3–75	4.7–6	4–7.6	2.5–5.9	8.56–11.98	329–341
Sorghum	11	70.7–72.97	3.23	1.97–6.7	1.6–1.7	6.07–11.16	329–339
Rice	4.99–6.94	74.3–82.86	1.90	1.63	0.99	11.6	369
Wheat	11.6–13.78	69.88–75.90	1.5–2.81	1.77	1.63	9.44	348–438

Table 1.
Nutritional analysis of different millets and other cereal grains (g per 100 g) [2, 5, 6, 10, 12–18].

frequently utilized as food and fodder in rural areas. With time millets are used to prepare a variety of value-added products, either as the base ingredient or by substituting any other grain. Despite not being a significant component of the average American or European diet, millets are increasingly valued as a component of multigrain and gluten-free cereal products in these regions. However, millet is a common staple food in many Asian and African nations, where it is also used to make a variety of ethnic cuisines and drinks [12]. The food products are made with a combination of wheat flour and other millet flour with millet flour to enhance the sensory profile. It is used in a variety of ways to get value-added products, either by substituting any other grain or as the base ingredient.

2. Nutritional, nutraceutical potential, and value-added products of different millets

2.1 Barnyard millet

2.1.1 Taxonomical classification of the Echinochloa frumentacea (common name—Indian Barnyard millet)

Kingdom	Order	Family	Genus	Species
•Plantae	•Poales	•Poaceae	•*Echinochloa*	•*frumentacea*

Echinochloa frumentacea is commonly known as *Indian Barnyard millet or billion dollar grass* (English); *Sanwa* (Hindi); *Shyama* (Sanskrit); *Oodalu* (Kannada); *Jhangora* (Uttarakhand); *Swank* (Punjabi); *Bhagar* (Maharashtra); *Samo or Morio* (Gujarati); *Kauda, Kautta, Kowda, Kowtta* (Malayalam); *Udali/Kodisama* (Telugu); *Shamula* (Bengali); and *Kuthiraivaali* (Tamil) [19, 20].

2.1.2 Nutritional properties of barnyard millet

Barnyard millet is a good source of protein, fat, carbohydrate and crude fiber, vitamins and micronutrients such as iron, zinc, calcium, magnesium, copper, and some essential amino acids, alkaloids, glycosides. It also contains phytochemicals such as phenolic acid, flavonoids, and tannins which serve as a good source of natural antioxidants [21, 22]. Barnyard millet is a rich source of several vitamins (thiamine, riboflavin, niacin, ascorbic acid, α-tocopherol), and contains some antinutrient constituents (a-amylase inhibitors, trypsin inhibitors, phytate, and tannins) [11, 22], fatty acids (linoleic acid, oleic, palmitic, stearic acid, and linolenic acid) [23]. Additionally it also contains amino acids like valine, isoleucine, leucine, lysine, phenylalanine, histidine, threonine, tryptophan, and methionine [15, 24].

2.1.3 Therapeutic attributes of barnyard millet

Barnyard millet is useful in the treatment of biliousness and constipation [25, 26] and allergic diseases such as atopic dermatitis [27]. It is best in lowering blood glucose and lipid levels therefore can be potentially recommended for patients with cardiovascular disease and diabetes mellitus [22]. It is also reported to have good antioxidant potential, anticarcinogenic, anti-inflammatory, antimicrobial, having a wound healing capacity [21]. Due to its high iron content, barnyard millet can help pregnant women who suffer from anaemia [6].

2.1.4 Food products prepared from barnyard millet

Barnyard millet is generally used in the preparation of different value-added products such as vermicelli, roti/chapati, noodles, biscuits, cookies, malt-based weaning food, extruded products, snack food, ladoo, halwa, biryani, dosa [6, 28–38].

2.2 Finger millet

2.2.1 Taxonomical classification of the Eleusine coracana (common name—Finger millet)

Kingdom	Order	Family	Genus	Species
•Plantae	•Poales	•Poaceae	•*Eleusine*	•*coracana*

Eleusine coracana is commonly known as *mandua, ragi, mandika* (Hindi); *marwa* (Bengali); *mandia* (Oriya); *nagli, bavto* (Gujarati); *keppagi, ragi, kelvaragu* (Tamil); *ragi chodi* (Telugu); *kaddo* (Nepal); *fingerhirse* (Germany); *petit mil, eleusine cultivee, coracan, koracan* (France); *bulo* (Uganda); *kambale, lupoko, mawele, amale, bule* (Zambia); *poko, rapoho, zviyo, njera, mazhovole* (Zimbabwe); *finger millet, African millet, koracan* (England); *dagussa, tokuso, barankiya, gadussa, dzoko* (Ethopia); *wimbi, mugimbi* (Kenya); *Mufhoho, mpogo* (South Africa); *Mwirubi, mbege, degi* (Tanzania); *Telebun, akima, bek, kal* (Sudan) [39–42].

2.2.2 Nutritional properties of finger millet

Finger millet is a very good source of micronutrients such as iron, zinc, calcium, potassium, selenium, copper, manganese, protein, fat, dietary fibers, polyphenols, pigments, phytates [22, 40, 43, 44], and amino acids such as lysine, valine, tryptophan, methionine, threonine, leucine, and isoleucine [40, 42, 45], vitamins (vitamin A, E, thiamine, riboflavin, niacin, folates, carotenoids) [40, 41, 43, 46], fatty acids (palmitic, oleic, linoleic, linolenic) [41]. Finger millet has the highest calcium content (344 mg/100 g) among all the food grains [1].

2.2.3 Therapeutic attributes of finger millet

Consumption of whole grain of finger millet is associated with health benefits, such as its hypoglycemic [42, 47], hypocholestrolemic characteristics, antitumerogenic, antidiarrheal, antiinflammatory, atherosclerogenic, antimicrobial, and antiulcerative properties [22, 42, 48, 49]. Additionally, it can reduce the risk of gastrointestinal malignancies, type II diabetes, cardiovascular diseases, and a variety of other illnesses [22]. Finger millet also helps in maintaining young and youthful skin. It is an excellent source of calcium and iron that helps to strengthen body bones and is a boon for anemic patients and also for those with low hemoglobin levels [43, 45].

2.2.4 Food products prepared from finger millet

Finger millet is generally used in the form of the whole meal for preparation of traditional foods, such as roti/chapati, mudde (dumpling), ambali (thin porridge), porridge, malting and weaning foods, papad, bakery products (biscuit, nank-hatai, muffins, rusk, cake, and bread), fermented foods (idli, dosa), dhokla, uthapam, chikki, beverages (chang/jnard, sur, madua, Themsing, rakshi, mingri, lohpani, koozh), pasta, vermicelli, noodles, ladoo, finger millet fritters, Vada, Soup, extruded products [1, 6, 22, 40, 42–44, 48, 50–52].

2.3 Pearl millet

2.3.1 Taxonomical classification of the Pennisetum glaucum (common name—Pearl millet)

Kingdom	Order	Family	Genus	Species
•Plantae	•Poales	•Poaceae	•*Pennisetum*	•*glaucum*

Pennisetum glaucum is commonly known as *spiked millet or pearl millet* (English); *bajra* (Bengali, Hindi, Oriya, Punjabi, Urdu); *bajree* (Rajasthani, Gujarati, Marathi); *sajje* (Kannada); *kambu* (Tamil); *sajja* (Telugu); *kambam* (Malyalam).

2.3.2 Nutritional properties of pearl millet

Pearl millet is significantly rich source of protein, fat, ash, soluble and insoluble dietary fibers, crude fiber, carbohydrate, resistant starch, minerals (calcium, iron, zinc, potassium, manganese, magnesium, copper) tannins, phytates, oxalates, flavanoids, vitamins (thiamine, riboflavin, niacin, vitamin E, A, folates), amino acids such as threonine, tryptophan, lysine, methionine, and fatty acids omega-9, omega-6, omega-3 [14, 53–56].

2.3.3 Therapeutic attributes of pearl millet

Pearl millet helps in reducing respiratory disease, migraine, and gall stones [57]. It is reported to have anticancerous, antidiabetic, antioxidant, antiinflammatory property [14, 55, 58]. It helps in increasing hemoglobin, antiallergic, helps in dealing with constipation, antiulcerative properties, helps in weight loss and essential for bone growth and development, prevents cardio-vascular diseases, regulates blood pressure, helpful in celiac disease [57, 58].

2.3.4 Food products prepared from pearl millet

Pearl millet is generally used in the preparation of various value-added food products such as cutlets, weaning food, vermicelli, instant beverage powder, biscuits, upma mixes, bread, cakes, muffins, roti/chapati, instant idli, dhokla, uthapam, kheer, extruded product, cookies, snack bar, beverages (appalu, Oshikundu) [6, 10, 32, 53, 54, 59–66].

2.4 Foxtail millet

2.4.1 Taxonomical classification of the Setaria italica (common name—Foxatil millet)

Kingdom	Order	Family	Genus	Species
•Plantae	•Poales	•Poaceae	•*Setaria*	•*italica*

Setaria italica is commonly known as *foxtail millet* (English); *kangni, rala* (Hindi); *navane* (Kannada); *korra* (Telugu); *thinai* (Malyalam, Tamil).

2.4.2 Nutritional properties of foxtail millet

Foxtail millet is found to be the good source of protein, crude fiber, fat, carbohydrates, vitamins (A, niacin, thiamine, riboflavin, E), fatty acids (palmitic acid, stearic acid, oleic acid, linoleic acid, linolenic acid, and arachidic acid), minerals (selenium, calcium, phosphorus, copper, magnesium, iron, potassium), amino acids (isoleucine, leucine, lysine, methionine, phenylalanine, threonine, valine, tryptophan). Additionally, it is abundant in phytochemicals such phytic acid [67–69], phenolic acid (chlorogenic acid, syringic acid, caffeic acid, ferulic and p-coumaric acid) [70].

2.4.3 Therapeutic attributes of foxtail millet

Foxtail millet is suitable for individuals suffering from diabetes mellitus due to its low glycemic index. It possess several health benefits like prevention of cancer, hypoglycemic, and hypolipidemic effects, curing dementia, helps in maintaining cholesterol level, antiproliferative activity, antilipidemic activity [6, 70], reduces inflammation, potentially promoting anticancer, antiaging, and improves the overall digestive health [68], increases kidney functionality, helps in development of body tissue and energy metabolism [71].

2.4.4 Food products prepared from foxtail millet

Foxtail millet is used as an important ingredient for preparing halwa, ladoo, noodles, soup, beverages (sikhye, yakju), cereal porridges, ready to eat extruded snacks, cookies, pancakes, cupcakes, biscuits, biryani, idli, dosa, dhokla, rice, upma, baby food, bread, korramurukulu, shakkarpara, ladoo [6, 10, 36, 38, 52, 70, 72–75].

2.5 Proso millet

2.5.1 Taxonomical classification of the Panicum miliaceum (common name—Proso millet)

Kingdom	Order	Family	Genus	Species
•Plantae	•Poales	•Poaceae	•*Panicum*	•*miliaceum*

Panicum miliaceum is commonly known as *proso millet* (English); *barri* (Hindi); *baragu* (Kannada); *varigulu* (Telugu); *panivaragu* (Tamil).

2.5.2 Nutritional properties of proso millet

Proso millet is said to be a rich source of crude fiber, carbohydrates, and protein. It is highly rich in amino acids (leucine, isoleucine, tryptophon, phenylalanine, valine, and methionine), minerals (zinc, copper, boron, phosphorus, potassium, iron, manganese). It is also rich in polyphenolic compounds such as phytic acid, catechin, tannin [76–81], vitamins like thiamine, riboflavin, folic acid, B6, niacin, vitamin E, and fatty acids (oleic, linoleic acid) [77, 81–83].

2.5.3 Therapeutic attributes of proso millet

Proso millet is potentially helpful in preventing cancer, heart disease and managing liver disease and diabetes [80]. Low glycemic index grains are beneficial for treating type 2 diabetes and cardiovascular conditions. It helps in reducing cholesterol and high blood pressure levesl and also helps in preventing cancer and reduces the risk of heart diseases, prevents liver injuries, celiac disease, and obesity. It also slows down the aging process and may protect against age-onset degenerative diseases [77, 78, 81].

2.5.4 Food products prepared from proso millet

Proso millet is generally used for preparing chakli, noodles, cookies, chapati, kitchari, chila, idli, dosa, namkeen, biscuits, halwa, Payasam, roti, bread, ready-to-eat breakfast cereal [32, 76, 79, 82, 84].

2.6 Little millet

2.6.1 Taxonomical classification of the Panicum sumatrense (common name—Little millet)

Kingdom	Order	Family	Genus	Species
•Plantae	•Poales	•Poaceae	•*Panicum*	•*sumatrense*

Panicum sumatrense is commonly known as *little millet* (English); *kutki* (Hindi); *same* (Kannada); *sama* (Telugu); *samai* (Tamil); *chama* (Malayalam).

2.6.2 Nutritional properties of little millet

Little millet is rich in protein, carbohydrates, minerals (iron, phosphorus, manganese, magnesium, iron, calcium, sodium, zinc), vitamins (thiamine, riboflavin, niacin, folic acid), and amino acids (tryptophan, phenylalanine, lysine, threonine, valine, leucine, isoleucine) [85, 86].

2.6.3 Therapeutic attributes of little millet

Little millet is helpful for diabetic patients, reduces blood glucose level, improves heart health, good for lowering cholesterol level, helps in weight loss [2, 71, 86].

It exhibits hypoglycemic and hypolipidemic effects [87] and prevents metabolic disorder [88].

2.6.4 Food products prepared from little millet

Little millet is generally used in the preparation of samaipayasam, roti, dosa, idli, pongal, khichadi, ladoo, chakli, bread, extruded snack, paddu, kheer, biscuits, pudding [6, 32, 85, 86, 89–93].

2.7 Kodo millet

2.7.1 Taxonomical classification of the Paspalum scrobiculatum (common name—Kodo millet)

Kingdom	Order	Family	Genus	Species
•Plantae	•Poales	•Poaceae	•*Paspalum*	•*scrobiculatum*

Paspalum scrobiculatum is commonly known as *kodo millet* (English); *koden, kodra* (Hindi); *harka* (Kannada); *arikelu* (Telugu); *varagu* (Tamil); *koovaragu* (Malayalam).

2.7.2 Nutritional properties of kodo millet

Kodo millet is a nutritious grain being rich in protein, crude fiber, carbohydrate, minerals (iron, calium, zinc, phosphorus, potassium, magnesium), vitamins like thiamine, riboflavin, niacin, folic acid; amino acids (lysine, threonine, valine, leucine, isoleucine) [6, 85, 86, 94]. It is also rich in polyphenolic compounds such as gallic acid, tannins, gentisic acid, protocatcchuic acid, caffeic acid, vanillic acid, syringic acid, ferulic acid, para coumaric acid, transcinnamic acid, and 5 *n*-alkyl-resorcinols [94].

2.7.3 Therapeutic attributes of kodo millet

Kodo millet has numerous health benefits like antidiabetic, antioxidant, antimicrobial, antiobesity, anticholesterol, antimutagenic, antioestrogenic, anticarcinogenic, antiinflammatory, antihypertension, and antiviral effects [95, 96]. It is useful in curing asthma, migraine, blood pressure, aging, heart attack, cardiovascular disease, and atherosclerosis [94].

2.7.4 Food products prepared from kodo millet

Kodo millet is used as an important ingredient for preparing roti/chapati, bread, cakes, extrusion of cereal-based products, gravy, soup, porridge, instant powders, dosa, ladoo, pudding, chakli, pongal, puttu, cutlet, methi rice, sev, idiyappam, cookies, kozhukattai, thattai, payasam, boli, sheera, pakoda, halwa, upma, idli, adai, murukkus, biscuits, yogurt, papad, thatuvadai, kolukattai, adhirasam, vermicelli, pasta [32, 37, 86, 89, 93, 95, 97, 98].

2.8 Sorghum

2.8.1 Taxonomical classification of the Sorghum bicolor (common name—Sorghum)

Kingdom	Order	Family	Genus	Species
•Plantae	•Poales	•Poaceae	•*Sorghum*	•*bicolor*

Sorghum bicolor is commonly known as *sorghum* (English); *jowar* (Hindi); *cholam* (Malyalam, Tamil); *jola* (Kannada); *Jwari* (Marathi); *Janha* (Oriya); *Juar* (Bengali, Gujarati); Jonnalu (Telugu); *guinea corn* (West Africa); *kaoliang* (China); *Kafir corn, milo* (United States).

2.8.2 Nutritional properties of sorghum

Sorghum is the good source of protein, crude fiber, fat, carbohydrates, fatty acids (linoleic, oleic, palmitic, stearic, linolenic), minerals (magnesium, sodium, selenium, phosphorus, potassium, calcium, iron, zinc, copper, and manganese), and amino acids (alanine, aspartic acid, glutamic acid, leucine, phenylalanine, proline, valine). Sorghum is a great source of bioactive substances such phenolic acids, flavanoids, tannins, 3-deoxyanthocyanidins, proanthocyanins, and carotenoids as well as vitamins B complex, D, E, and K, as well as β-carotene [17, 18, 99–106].

2.8.3 Therapeutic attributes of sorghum

Sorghum is beneficial in curing diseases such as obesity, diabetes, celiac disease, dyslipidemia, cardiovascular disease, cancer, heart diseases, dyslipidemia, maintains cholesterol level, bone health, hypertension, and prevents anemia. The grain also has antioxidant, anti-inflammatory, antimicrobial, antithrombotic, and anticancer activity [18, 102–104, 106–110].

2.8.4 Food products prepared from sorghum

Sorghum is used as an important ingredient for the preparation of gluten-free breads, cookies, porridge (madeda, logma or asida), tortillas, enjera, pasta, muffins, nefro, cupcakes, kisra, ladoo, chapati, chaat, beverages (tela, arekie, pito, bouza, kunu-zaki, mahewu, tea, chibuku, bushera, burukutu, ikigage, enturire, merissa, otika, thobwa, Orubisi), juice, cake, noodles, extruded products, thekua, tender sweet sorghum (hurda) [17, 103, 105, 107, 109, 111–118].

3. Conclusion

Millets play an important role in the food security and generating economy of developing countries throughout the world. In this chapter, we have focused on the nutritional profile and medicinal properties of different millet grains and their value-added food products. According to the findings of research, millet grains are equivalent to major grains in terms of the amount of health-promoting nutrients they contain, including phenolic compounds, dietary fiber, minerals, and vitamins. They also offer a number of other possible health benefits. It will be beneficial to encourage the use of millet grains in urban areas in order to open new markets for farmers and improve their income if millet food products that offer convenience, taste, texture, color, and shelf-stability are made popular in society. Promoting growth and development and marketing might help safeguard food and nutrition, create jobs, and generate revenues. These healthy "super foods," as the new health industry christened, have become a part of the modern daily diet of health-conscious people.

Author details

Smita Rana[1*] and Narendra Singh Bhandari[2]

1 Kumaun University, Nainital, Uttarakhand, India

2 S.S.J. University, Almora, Uttarakhand, India

*Address all correspondence to: smitarana2@gmail.com

References

[1] Chandra A, Singh AK, Mahto B. Processing and value addition of finger millet to achieve nutritional and financial security. International Journal of Current Microbiology and Applied Sciences. 2018;7:2901-2910

[2] Ambati K, Sucharitha KV. Millets—Review on nutritional profiles and health benefits. International Journal of Recent Scientific Research. 2019;**10**:30693-30695

[3] Kumar BG, Sivaramane N, Rao CS. Economic analysis of production and consumption of finger millet in India. Multilogic in Science. 2020;**10**:1-7

[4] Kumari P, Kumar A. Role of millets in ensuring food and nutritional security. European Journal of Molecular & Clinical Medicine. 2020;7:2544-2551

[5] Nithiyanantham S, Kalaiselvi P, Mahomoodally MF, et al. Nutritional and functional roles of millets—A review. Journal of Food Biochemistry. 2019;**43**:1-10

[6] Kumar A, Tomer V, Kaur A, et al. Millets: A solution to agrarian and nutritional challenges. Agriculture & Food Security. 2018;7:1-15

[7] Chamoli V, Badoni A, Bahuguna N, et al. Finger millet (*Eluesine coracana*): Nutritional status, health benefits and processing status—A review. Journal of Pharmacognosy and Phytochemistry. 2018;7:80-83

[8] FAOSTAT. Production-Yield Quantities of Millets in World + (total) 1994-2020. 2022. Available from: https://www.fao.org/faostat/en/#data/QCL/visualize [Accessed: November 18, 2022]

[9] Ransumithila C, Saravakumar R. Development of value added millet based nutritious "Instant Dhokla Mix". International Journal of Chemical Studies. 2019;7:4878-4882

[10] Mishra P, Prakash HG, Devi S, et al. Nutritional quality of millets and their value added products with the potential health benefits: A review. International Journal of Current Microbiology and Applied Sciences. 2021;**10**:163-175

[11] Saleh ASM, Zhang Q, Chen J, et al. Millet grains: Nutritional quality, processing, and potential health benefits. Comprehensive Reviews in Food Science and Food Safety. 2013;**12**:281-295

[12] Jaybhaye RV, Pardeshi IL, Vengaiah PC, et al. Processing and technology for millet based food products: A review nutrient composition of millets. Journal of Ready to Eat Foods. 2014;**1**:32-48

[13] Himanshu K, Chauhan M, Sonawane SK, et al. Nutritional and nutraceutical properties of millets. Journal of Nutrition & Dietetics. 2018;**1**:1-10

[14] Hassan ZM, Sebola NA, Mabelebele M. The nutritional use of millet grain for food and feed: A review. Agriculture & Food Security. 2021;**10**:1-14

[15] Vanga SK, Singh A, Orsat V, et al. Final Technical Report/Rapport Technique Final Annex. 2.5: Nutritional Comparison of Millets With Other Superfoods, 2018

[16] Hymavathi TV, Roberts TPP, Jyothsna E, et al. Proximate and mineral content of ready to use minor millets. International Journal of Chemical Studies. 2020;**8**:2120-2123

[17] Adebo OA. African sorghum-based fermented foods: Past, current and future prospects. Nutrients. 2020;**12**:1-25

[18] Chhikara N, Abdulahi B, Munezero C, et al. Exploring the nutritional and phytochemical potential of sorghum in food processing for food security. Nutrition & Food Science. 2018;**49**:318-332

[19] Ugare R, Chimmad B, Naik R, et al. Glycemic Index and Significance of Barnyard Millet (*Echinochloa frumentacae*) in Type II Diabetics. Journal of Food Science and Technology. 2011;**51**:392-395

[20] Kaur H, Sharma S. An overview of Barnyard Millet (*Echinochloa frumentacea*). Journal of Pharmacognosy and Phytochemistry. 2020;**9**:819-822

[21] Renganathan VG, Vanniarajan C, Karthikeyan A, et al. Barnyard millet for food and nutritional security: Current status and future research direction. Frontiers in Genetics. 2020;**11**:1-22

[22] Panwar P, Dubey A, Verma AK. Evaluation of nutraceutical and antinutritional properties in barnyard and finger millet varieties grown in Himalayan region. Journal of Food Science and Technology. 2016;**53**:2779-2787

[23] Bora P, Ragaee S, Marcone M. Characterisation of several types of millets as functional food ingredients. International Journal of Food Sciences and Nutrition. 2019;**70**:714-724

[24] Geervani P, Eggum BO. Nutrient composition and protein quality of minor millets. Plant Foods for Human Nutrition. 1989;**39**:201-208

[25] Srinivasan A, Ekambaram PS, Perumal SS, et al. Chemical characterization and immunostimulatory activity of phenolic acid bound arabinoxylans derived from foxtail and barnyard millets. Journal of Food Biochemistry. 2019;**44**:1-9

[26] Chopra RN, Nayar SL, Chopra IC. Glossary of Indian Medicinal Plants (Including the Supplement). New Delhi, India: Council of Scientific and Industrial Research (CSIR); 1986

[27] Watanabe M. Antioxidative phenolic compounds from Japanese barnyard millet (*Echinochloa utilis*) grains. Journal of Agricultural and Food Chemistry. 1999;**47**:4500-4505

[28] Chandraprabha S, Sharon CL, Panjikkaran ST, et al. Development and nutritional qualities of vermicelli prepared from barnyard millet and *Ekanayakam* root bark. Journal of Pharmacognosy and Phytochemistry. 2017;**6**:2359-2362

[29] Dhumal CV, Pardeshi IL, Sutar PP, et al. Development of potato and barnyard millet based ready to eat (RTE) fasting food. Journal of Ready to Eat Foods. 2014;**1**:11-17

[30] Goel K, Goomer S, Aggarwal D. Formulation and optimization of value-added barnyard millet vermicelli using response surface methodology. Asian Journal of Diary and Food Research. 2021;**40**:55-61

[31] Jaybhaye RV, Srivastav PP. Development of barnyard millet ready-to-eat snack food: Part II. Food Science. 2015;**6**:285-291

[32] Reddy M, Shivakumara CS, Aneesha. Flour and dough quality of milets and their suitability for preparation of traditional south Indian roti. Journal of Clinical and Biomedical Sciences. 2019;**9**:13-18

[33] Surekha N, Naik RS, Mythri S, et al. Barnyard Millet (*Echinochloa frumentacea* Link) cookies: Development, value addition, consumer acceptability, nutritional and shelf life evaluation. Journal of Environmental Science, Toxicology and Food Technology. 2013;**7**:1-10

[34] Surekha N, Devi R, Naik RS. Development of value added low glycemic index barnyard millet (*Echinochloa frumentacea* Link) noodles. International Journal of Food and Nutritional Sciences. 2013;**2**:20-24

[35] Thathola A, Srivastava S. Physicochemical properties and nutritional traits of millet-based weaning food suitable for infants of the Kumaon Hills, Northern India. Asia Pacific Journal of Clinical Nutrition. 2002;**11**:28-32

[36] Verma S, Srivastava S, Tiwari N. Comparative study on nutritional and sensory quality of barnyard and foxtail millet food products with traditional rice products. Journal of Food Science and Technology. 2014;**52**:1-9

[37] Vijayakumar PT, Mohankumar JB. Quality evaluation of dosa from millet flour blend incorporated composite flour. Food Science. 2011;**34**:2624-2629

[38] Anju T, Sarita S. Suitability of foxtail millet (*Setaria italica*) and barnyard millet (*Echinochloa frumentacea*) for development of low glycemic index biscuits. Malaysian Journal of Nutrition. 2010;**16**:361-368

[39] Ramashia SE, Gwata ET, Meddows-Taylor S, et al. Nutritional composition of fortified finger millet (*Eleusine coracana*) flours fortified with vitamin B2 and zinc oxide. Food Research. 2021;**5**:456-467

[40] Arjun D, Kumar R, Singh C. Finger millet for nutritional security and source of food. Advance Journal of Food Science and Technology. 2014;**2**:184-190

[41] Ramashia SE, Anyasi TA, Gwata ET, et al. Processing, nutritional composition and health benefits of finger millet Sub-Saharan Africa. Food Science and Technology. 2019;**39**:253-266

[42] Ambre PK, Sawant AA, Sawant PS. Processing and value addition: A finger millet review. Journal of Pharmacognosy and Phytochemistry. 2020;**9**:375-380

[43] Rathore T, Singh R, Kamble DB, et al. Review on finger millet: Processing and value addition. Pharma Innovation Journal. 2019;**8**:283-291

[44] Ambre PK. Processing and value addition: A finger millet review. Vigyan Varta. 2021;**2**:42-52

[45] Devi VN, Sinthiya R. Development of value added product from finger millet (*Eleusine coracana*). International Journal of Research—Granthaalayah. 2018;**6**:109-119

[46] Cholo A. Evaluation and demonstration of new food product of enset, orange fleshed sweet potato and finger millet blends for alleviation of nutritional insecurity. Journal of Food Technology and Preservation. 2020;**4**:1-5

[47] Kumari LP, Sumathi S. Effect of consumption of finger millet on hyperglycemia in non-insulin dependent diabetes mellitus (NIDDM) subjects. Plant Foods for Human Nutrition. 2002;**57**:205-213

[48] Sharma D, Yamer P. Value addition of products from finger millet (*Eleusine coracana*). Journal of Postharvest Technology. 2022;**10**:120-138

[49] Chandra D, Chandra S, Pallavi, et al. Finger millet (*Eleusine coracana*

(L.) Gaertn): A power house of health benefiting nutrients, a review. Food Science and Human Wellness 2016;5: 1-19

[50] Gupta SM, Arora S, Mirza N, et al. Finger millet: A "certain" crop for an "uncertain" future and a solution to food insecurity and hidden hunger under stressful environments. Frontiers in Plant Science. 2017;**8**:1-11

[51] Verma V, Patel S. Value added products from nutri-cereals: Finger millet (*Eleusine coracana*). Emirates Journal of Food and Agriculture. 2013;**25**:169-176

[52] Ayin P, Gupta A, Verma T, et al. Organoleptic and nutritional evaluation of value added food products incorporated Houttuynia Cordata, figer millet and foxtail millet. International Journal of Home Science. 2021;7:167-171

[53] Mishra S, Mishra S. Nutritional analysis of value-added product by using pearl millet, quinoa and prepare ready-to-use upma mixes. International Journal of Food Sciences and Nutrition. 2016;**1**:41-43

[54] Singh R, Singh K, Nain MS. Nutritional evaluation and storage stability of popped pearl millet bar. Current Science. 2021;**120**:1374-1381

[55] Dube M, Nyoni N, Bhebhe S, et al. Pearl millet as a sustainable alternative cereal for novel value-added products in Sub-Saharan Africa: A review. Agricultural Reviews. 2021;**42**:1-4

[56] Ragaee S, Abdel-Aal ESM, Noaman M. Antioxidant activity and nutrient composition of selected cereals for food use. Food Chemistry. 2006;**98**:32-38

[57] Malik S. Pearl millet-nutritional value and medicinal uses. International Journal of Advance Research and Innovative Ideas in Education. 2015;**1**:414-418

[58] Pei JJ, Umapathy VR, Vengadassalapathy S, et al. A review of the potential consequences of pearl millet (*Pennisetum glaucum*) for diabetes mellitus and other biomedical applications. Nutrients. 2022;**14**:1-12

[59] Gupta P, Singh Y. Development and sensory evaluation of pearl millet based value added biscuits blended with different non-wheat flours. Asian Pacific Journal of Health Sciences. 2020;7:4-7

[60] Kale PG, Vairagar PR. Application of pearl millet in functional food. International Journal of Agricultural Engineering. 2018;**11**:90-94

[61] Obilana AO. Nutritional, Physico-Chemical and Sensory Characteristics of a Pearl Millet-Based Instant Beverage Powder. 2013. Available from: http://ir.dut.ac.za/handle/10321/1145

[62] Balasubramanian S, Yadav DN, Kaur J, et al. Development and shelf-life evaluation of pearl millet based upma dry mix. Journal of Food Science and Technology. 2012;**51**:1-8

[63] Sumathi A, Ushakumari SR, Malleshi NG. Physico-chemical characteristics, nutritional quality and shelf-life of pearl millet based extrusion cooked supplementary foods. International Journal of Food Sciences and Nutrition. 2007;**58**:350-362

[64] Gupta A, Singh R, Prasad R, et al. Nutritional composition and polyphenol content of food products enriched with millets and drumstick leaves. International Journal of Food and Fermentation Technology. 2017;7:337-342

[65] Suma P F, Urooj A, MR A, et al. Sensory, physical and nutritional qualities of cookies prepared from pearl millet (*Pennisetum typhoideum*). Journal of Food Processing & Technology. 2014;**5**:1-6

[66] Jha A, Tripathi AD, Alam T, et al. Process optimization for manufacture of pearl millet-based dairy dessert by using response surface methodology (RSM). Journal of Food Science and Technology. 2011;**50**:367-373

[67] Mohamed TK, Issoufou A, Zhou H. Antioxidant activity of fractionated foxtail millet protein hydrolysate. International Food Research Journal. 2012;**19**:207-213

[68] Verma KC, Joshi N, Rana AS, et al. Quality parameters and medicinal uses of foxtail millet (*Setaria italica* L.): A review. Journal of Pharmacognosy and Phytochemistry. 2020;**9**:1036-1038

[69] He L, Zhang B, Wang X, et al. Foxtail millet: Nutritional and eating quality, and prospects for genetic improvement. Frontiers of Agricultural Science and Engineering. 2015;**2**:124-133

[70] Sharma N, Niranjan K. Foxtail millet: Properties, processing, health benefits, and uses. Food Review International. 2018;**34**:329-363

[71] Wandhekar SS, Sadawarte SK, Pawar VS, et al. Production status, nutritional aspects and health benefits of millets—A review. Journal of Emerging Technologies and Innovative Research. 2021;**10**:217-224

[72] Doddamani S, Yenagi NB. Nutrient composition of pre-treated foxtail millet rice. International Journal of Current Microbiology and Applied Sciences. 2018;**7**:1314-1322

[73] Thara KD, Nazni P. Formulation and quality evaluation of foxtail millet and semolina incorporated ready-to-cook upma mix. Bioscience Biotechnology Research Communications. 2021;**14**:1531-1537

[74] Mounika P, Gnanashree G, Sangeetha P, et al. Effect of incorporation of foxtail millet for preparation of dosa breakfast mix and comparison with the traditional dosa mix. Biological Forum—An International Journal. 2021;**13**:501-507

[75] Sadhu VJ, Naidu S. A study on functional and nutritional characteristics of barnyard millet and foxtail millet. Food and Nutrition Journal. 2021;**6**:1-10

[76] Karkannavar SJ, Hegde SC, Desai SR, et al. Standardization and nutrient composition of the proso millet chakli. Biological Forum—An International Journal. 2021;**13**:44-50

[77] Habiyaremye C, Matanguihan JB, D'Alpoim Guedes J, et al. Proso millet (*Panicum miliaceum* L.) and its potential for cultivation in the Pacific Northwest, U.S.: A review. Frontiers in Plant Science. 2017;**7**:1-17

[78] Karkannavar SJ, Shigihalli S, Nayak G, et al. Physico-chemical and nutritional composition of proso millet varieties. Pharma Innovation Journal. 2021;**10**:136-140

[79] Sandhvi S, Singh V. Develop the value added products and sensory evaluation of proso millet fresh products. International Journal of Current Microbiology and Applied Sciences. 2021;**10**:262-267

[80] Bhat S, Nandini C, Srinathareddy S, et al. Proso millet (*Panicum miliaceum* L.)—a climate resilient crop for food and nutritional security: A review. Environment Conservation Journal. 2019;**20**:113-124

[81] Agarwal S, Chauhan ES. The nutritional composition, various processing and health benefits of proso millet: A review. Research Journal of Pharmacy and Technology. 2019;**12**:4013-4017

[82] Kalinová J. Nutritionally important components of proso millet (*Panicum miliaceum* L.). Food. 2007;**1**:91-100

[83] Kulkarni DB, Sakhale BK, Giri NA. A potential review on millet grain processing. International Journal of Nutritional Sciences. 2018;**3**:1-8

[84] Sandhvi S, Singh V. Standardization, nutritional and sensory analysis of Proso Millet Halwa. Pharma Innovation Journal. 2022;**11**:2250-2252

[85] Dey S, Saxena A, Kumar Y, et al. Understanding the antinutritional factors and bioactive compounds of kodo millet (*Paspalum scrobiculatum*) and little millet (*Panicum sumatrense*). Journal of Food Quality. 2022;**2022**:1-19

[86] Bhat S, Nandini C, Tippeswamy V, et al. Significance of small millets in nutrition and health: A review. Asian Journal of Diary and Food Research. 2017;**36**:1-6

[87] Patil KB, Chimmad BV, Itagi S. Glycemic index and quality evaluation of little millet (*Panicum miliare*) flakes with enhanced shelf life. Journal of Food Science and Technology. 2015;**52**:6078-6082

[88] Kumari P, Singh YV, Yadav R, et al. Analytical study of little millets and therapeutic effect of little millets based on its components in type 2 diabetes mellitus. International Journal of Herbal Medicine. 2022;**10**:11-14

[89] Chouhan RS, Niranjan HK, Sharma HO. Economics of value added products of kodo and kutki. Indian Journal of Economics and Development. 2019;**15**:465-469

[90] Vasure A, Hamad R, Wakde D, et al. Nutritional properties of extruded prepared by little millet (Kutki) and defatted soya flour. Indian Journal of Pure & Applied Biosciences. 2017;**5**:749-754

[91] Madalageri DM, Yenagi NB, Shirnalli G. Evaluation of little millet Paddu for physico-chemical nutritional, microbiological and sensory attributes. Asian Journal of Diary and Food Research. 2016;**35**:58-64

[92] Shah KN, Pawar PA. Development and sensory evaluation of Kheer prepared from little millet (Kutki), milk and honey from Melghat. International Journal of Food Sciences and Nutrition. 2021;**6**:51-56

[93] Kaushik N, Chauhan K, Aggarwal M, et al. State-of-the-art of knowledge on underutilized millets: Kodo and kutki, grown in tribal areas of India. International Journal of Agricultural Science and Research. 2022;**12**:35-56

[94] Bunkar DS, Goyal SK, Meena KK, et al. Nutritional, functional role of kodo millet and its processing: A review. International Journal of Current Microbiology and Applied Sciences. 2021;**10**:1972-1985

[95] Kumari P, Nazni P. Formulation of value added yogurt prepared with kodo millet milk. Journal of Nutrition and Dietetics. 2021;**4**:1-4

[96] Sharma S, Sharma N, Handa S, et al. Evaluation of health potential of nutritionally enriched Kodo millet (*Eleusine coracana*) grown in Himachal Pradesh, India. Food Chemistry. 2017;**214**:162-168

[97] Muragod PP, Muruli NV, Padeppagol S, et al. Develop the value added products and evaluate the storage quality of kodo millet grains products. Indian Journal of Pure & Applied Biosciences. 2019;**7**:97-107

[98] Deshpande SS, Mohapatra D, Tripathi MK, et al. Kodo millet-nutritional value and utilization in indian foods. Journal of Grain Processing and Storage. 2015;**2**:16-23

[99] Mohammed ZS, Mabudi AH, Murtala Y, et al. Nutritional analysis of three commonly consumed varieties of sorghum (*Sorghum bicolor* L.) in Bauchi State, Nigeria. Journal of Applied Sciences and Environmental Management. 2019;**23**:1329-1334

[100] Zarei M, Amirkolaei AK, Trushenski JT, et al. Sorghum as a potential valuable aquafeed ingredient: Nutritional quality and digestibility. Agriculture. 2022;**12**:1-17

[101] Tasie MM, Gebreyes BG. Characterization of nutritional, antinutritional, and mineral contents of thirty-five sorghum varieties grown in Ethiopia. International Journal of Food Science. 2020;**2020**:1-11

[102] Pontieri P, Troisi J, Calcagnile M, et al. Chemical composition, fatty acid and mineral content of food-grade white, red and black sorghum varieties grown in the mediterranean environment. Food. 2022;**11**:1-11

[103] Rashwan AK, Yones HA, Karim N, et al. Potential processing technologies for developing sorghum-based food products: An update and comprehensive review. Trends in Food Science and Technology. 2021;**110**:168-182

[104] Mohamed HI, Fawzi EM, Basit A, et al. Sorghum: Nutritional factors, bioactive compounds, pharmaceutical and application in food systems: A review. Phyton-International Journal of Experimental Botany. 2022;**91**:1-24

[105] Khalid W, Ali A, Arshad MS, et al. Nutrients and bioactive compounds of *Sorghum bicolor* L. used to prepare functional foods: A review on the efficacy against different chronic disorders. International Journal of Food Properties. 2022;**25**:1045-1062

[106] Xiong Y, Zhang P, Warner RD, et al. Sorghum grain: From genotype, nutrition, and phenolic profile to its health benefits and food applications. Comprehensive Reviews in Food Science and Food Safety. 2019;**00**:1-22

[107] Uñate-Fraga S, García-López JI, Flores-Naveda A, et al. Grain yield, nutritional, polyphenols and antioxidant capacity in accessions of sorghum (*Sorghum bicolor* L. Moench). Notulae Botanicae Horti Agrobotanici Cluj-Napoca. 2022;**50**:1-20

[108] Birhanu S. Potential benefits of sorghum [*Sorghum bicolor* (L.) Moench] on human health: A review. International Journal of Food Engineering and Technology. 2021;**5**:8-18

[109] Samarth D, More D, Dipak S. Technology for tender sweet sorghum (Hurda) production and its health benefits: A review. International Journal of Food Sciences and Nutrition. 2017;**2**:188-191

[110] de Morais CL, Pinheiro SS, Martino HSD, et al. Sorghum (*Sorghum bicolor* L.): Nutrients, bioactive compounds, and potential impact on human health. Critical Reviews in Food Science and Nutrition. 2017;**57**:372-390

[111] Singh V, Sati V, Agarwal S. Potential of germinated sorghum (*Sorghum bicolor*) for utilization as a health food. International Journal of Science, Environment. 2016;**5**:4203-4212

[112] Deribe Y, Kassa E. Value creation and sorghum-based products: What synergetic actions are needed? Cogent Food & Agriculture. 2020;**6**:1-16

[113] Hu W, Zhou L, Chen J, hong. Conversion sweet sorghum biomass to produce value-added products. Biotechnology for Biofuels and Bioproducts. 2022;**15**:1-11

[114] Sikandra, Boora P. Nutritional evaluation of sorghum and chickpea incorporated value added products. Journal of Dairying, Foods and Home Sciences. 2009;**28**:181-185

[115] Alisha, Dubey RP. Preparation of value added healthy food product (Thekua) with incorporation of sorghum millet. International Journal of Food Sciences and Nutrition. 2018;**3**:131-136

[116] Weerasooriya DK, Bean SR, Nugusu Y, et al. The effect of genotype and traditional food processing methods on in vitro protein digestibility and micronutrient profile of sorghum cooked products. PLoS One. 2018;**13**:1-22

[117] Ahmed SER, Ahmed WAEA, Mohamed MA, et al. Fortified sorghum as a potential for food security in rural areas by adaptation of technology and innovation in Sudan. Journal of Food, Nutrition and Population Health. 2017;**1**:1-17

[118] Tegeye M, Kaur A, Kaur J, et al. Value added convenience food from composite sorghum-maize-sweet potato flour blends. Indian Journal of Agricultural Sciences. 2019;**89**:1906-1910

Chapter 3

Finger Millet Scientific Cultivation and Its Uses in India

Brijesh Kumar, Magan Singh, Avaneesh Kumar, Deepak Kumar, Ankit Kumar Singh, Rakesh Kumar and Sandeep Kumar

Abstract

Finger millet (*Eleusine coracana* L.) is a prevalent grain crop in the dry parts of Asia and Africa. It prolongs to be a staple food and is known locally as "Ragi" in southern states like Karnataka, Andhra Pradesh, and Telangana, especially in Karnataka. It serves as fodder as well as grain. Its grain is the richest source of calcium and it is utilized in a wide range of food products, including cakes, puddings, sweets, and other baked products. There are many homemade products prepared with finger millet and some of the well-known products are ragi roti, ragi dosa, ragi balls, ragi porridge, ragi upma, ragi cakes, and ragi biscuits. In addition, it is used to manufacture beer and liquor (known as arake or areki in Ethiopia), and also its different products are fed to animals. Due to its high fiber, mineral, vitamin, macro and micro-nutrient, phytochemicals contents, and its endowing ability to fight off chronic diseases. A cheap, satisfying, and healthful diet can be made by including finger millet in a regular diet.

Keywords: finger millet, cultivation, nutritional importance, quality, uses

1. Introduction

Finger millet (*Eleusine coracana*) is a domesticated crop of African origin and it is spread throughout the world. It is also known as korakan, ragi, or dagusa in India. Even though the wild progenitor species (*Eleusine africana*) is well recognized, this species was first domesticated in Africa [1]. Ecologically, it is thought to originate from a highland region as a crop and is frequently planted in hilly regions [2]. The crop is grown in a wide range of challenging environmental situations. Finger millet blessing for vast arid and semi-arid regions because it can be cultivated on low-fertility soils [3]. Finger millet is clearly regarded as a staple food and it is used as animal feed (straw) in industrialized countries and as food (grains) in underdeveloped [4]. It is a major crop of semi-arid and arid regions and developing nations of Asia and Africa [5]. Finger millet is the fourth most produced millet in the world, behind sorghum (*Sorghum bicolor*), pearl millet (*Pennisetum glaucum*), and foxtail millet (*Setaria italica*) [6]. Millet grains are rich in vitamins, iron, carbohydrates, calcium (Ca), potassium, zinc, phosphorus, magnesium, and vital amino acids, they are also nutritionally superior to rice and wheat [7]. The seed coat (testa), embryo, and endosperm are the three primary components of the finger millet grain. Finger millet grain are red, yellow, white, tan, red, brown, and violet, in color and this type of variety grown around the world. It is compared to other millets including foxtail millet, pearl millet, Kodo millet, and proso millet, finger millet distinctive due to the existence of five layered testa. This might be one of the reasons why finger millet has a greater level of dietary fiber [8]. Finger millet's ability to withstand droughts may be linked to its high antioxidant content. In underdeveloped countries, there is huge potential for millet grains to be transformed into foods and beverages with additional value [9].

2. Nutritional importance of finger millet

Next to fine cereal grains [10] and gluten-free cereal goods [7], finger millets are a prominent diet in Asia and Africa's resource-poor nations since they account for 75% of total caloric consumption. Because of the high amount of dietary fiber, iron, zinc, calcium, phosphorus, potassium, vitamin B, and essential amino acids, it is nutritionally superior to wheat and rice [11, 12]. In comparison to other millets, finger millet has higher levels of lysine, threonine, and valine [13, 14]. It contains some of the anti-nutrients *viz*., Phytates, polyphenols, tannins, trypsin dietary fiber, and inhibitory substances. Dietary fiber and polyphenols in finger millet have a number of health advantages, including actions that are anti-diabetic, antioxidant, hypocholesterolemic, anti-microbial, delay nutrient absorption, raise fecal bulk, and lower blood lipid levels [15]. It was discovered that finger millet contains more than 50 phenolic compounds from a variety of groups, including phenolic acids and their derivatives, dehydrodiferulates, dehydrotriferulates, flavan-3-ol monomers and dimers, flavonols, flavones, and flavanonols [7]. Furthermore, finger millet is helpful in managing a number of physiological conditions, including diabetes mellitus, hypertension, vascular fragility, hypercholesterolemia, and the prevention of low-density lipoprotein (LDL) oxidation [16]. The risk of cardiovascular diseases, type II diabetes, gastrointestinal malignancies, and other health problems can be reduced by consuming finger millet and its products on a daily basis [17].

3. Status area and production

The majority of finger millet is grown in semi-arid tropical regions of Asia and Africa. The southern states of India are where finger millet is primarily grown in Asia, and these regions have ideal growing conditions. Finger millet is the fourth most produced millet in the world after sorghum, pearl millet (*Cenchrus americanus*), and foxtail millet (*Setaria italica*). India produces 1.70 million tonnes of finger millet, which is grown over 1.07 million hectares' area [18]. The states Karnataka, Andhra Pradesh, Tamil Nadu, Kerala, Telangana, Uttarakhand, Jharkhand, Madhya Pradesh, and Haryana are important producers of finger millet. In addition to being utilized for direct human consumption, finger millet has a wide range of other uses, including animal feed, distilleries, and food processing for value-added goods [19].

4. Seed and sowing methods

4.1 Land preparation

One deep plowing is advised with a mold-board plow in the months of April or May. Before sowing, smooth seedbed preparation with secondary tillage using a cultivator and multiple tooth hoe is required. Therefore, appropriate seed and land preparation aid in better germination, reduce weed issues and effectively save soil moisture. Small seeds take about 5–7 days to germinate. In Uttaranchal, where it is difficult to carry out frequent plowing operations, efficient soil turning and digging, weed removal from perennial plants, land smoothening, and the provision of an inward slope with a shallow drain all help in draining surplus rainwater [20].

4.2 Seed rate and sowing

The recommended seed rate for sowing a one-hectare field is 8–10 kg of seed and 4–5 kg of seed sufficient for transplanting. To avoid infections, seeds should be treated with Thiram @ 2.5 g/kg of seeds. The best time for sowing is between June and July for the *Kharif* and September and October for the rabi. Usually, crops are grown during the *Kharif* season. In some areas, the crop is cultivated in irrigated conditions during the rabi season.

The line sowing is advantageous because it promotes inter-cultivation and provides efficient weed control and sowing with a seed drill and keeping space 22.5–30.0 cm between rows and 7.5–10.0 cm between plants, it is help in to maintain an ideal plant population of 4–5 lakh/ha (**Tables 1** and **2**) [20, 21].

Name of the state	Sowing duration
Andhra Pradesh	Early *kharif* (May–August)
	Late *kharif* (July–November)
	Late Rabi (November–March)
Bihar, Jharkhand, Gujarat, Himachal Pradesh, Uttar Pradesh, Uttara Khand, Orissa	*Kharif* (June–September)
Karnataka, Tamil Nadu	Kar or Early *Kharif* (April–August)
	Kharif (July–November)
	Rabi (October–September)
	Summer (January–May)

Table 1.
Finger millet is grown in all the cropping seasons in different parts of the country.

5. Cropping systems

State	Systems
Karnataka, Tamil Nadu and Andra Pradesh	Finger millet + pigeon pea (8–10:2)
	Finger millet +filed bean (8:1)
	Finger millet + soybean (4:1)
Bihar	Finger millet + pigeon pea (6:2)
Uttaranchal	Finger millet and soybean mixed together in 90:10 per cent proportion by weight basis
North hilly areas	Finger millet + soybean in *Kharif* and oats in rabi is an ideal remunerative sequence
Maharashtra (Kolhapur)	Finger millet + black gram/moong bean (6–8: 1) (Sub montane regions)

Table 2.
Intercropping systems of finger millet growing in different regions and states of India [20, 21].

6. Nutrient management

The application of farm yard manure before sowing of the crop is 5–10 t/ha. Fertilizer applications give better results to produce a higher yield. Fertilizer recommendations for finger millet are 40 kg nitrogen (N), 20 kg phosphorus (P), and 20 kg potassium(K) per ha for rainfed cultivation and 60 kg N, 30 kg P, and 30 kg K per ha for irrigation. Rao et al. [22] found that application rates of up to 40 kg N/ha of nitrogen fertilizer resulted in improved grain yield and protein content in finger millet. According to Hedge and Gowda [23] found that the application rate of phosphorus was increased from 30 to 60 kg/ha phosphorus to improve grain yield. The application of zinc (Zn), boron (B), and sulfur (S) coupled with N, P increased the yield of finger millet grains (56%), Stover biomass (44%), total biomass (48%), and plant uptake of Zn (66%) and B (22%) compared to the addition of N and P alone, according to Srinivasarao et al. [24]. Rao et al. [25] reported that the application of B and Zn along with N, P, and S fertilizer increased finger millet's grain, straw yield, and nutrient uptake of N, P, S, B, and Zn. Finger millet treated with *Azospirillium + arbuscular mycorrhizal* (AM) fungi + PSB improved plant growth and N, P uptake [26]. When finger millet was treated with AM fungus in comparison to uninoculated plants the absorption of macronutrients (N, P) and micronutrients (Zn, Cu) was improved in plants [27].

The application of *Azotobacter* and PSB coupled with fertilizer (based on soil testing) and FYM (10 t/ha) enhanced finger millet yield in comparison to the recommended fertilizer treatment alone [28].

7. Water management

Finger millet cultivated in rainfed conditions does not require any irrigation, but if rain delays for a long period of time between tillering and flowering stages, irrigation should be necessary to produce a reasonable yield. To fulfill the water requirement of crops to create furrows and ridges for irrigation [29]. Because the crop cannot thrive in wet conditions, it is very crucial to remove excess water after rain. Since it is a drought-tolerant crop, additional watering at critical stages of the growth cycle results in a good yield from finger millet-based farming systems [30]. If we want to get a good yield from a finger millet-based cropping system in a drought-prone area, drip irrigation is an alternate and efficient technique of irrigation. Other management techniques, including as mulching, intercropping, growing pulses in intercropping, Intercultivation, and application of organic manure, boost soil moisture retention and increase the yield of finger millet-based cropping systems. Finger millet and pulse intercropping also reduce soil erosion and the loss of nutrients from the top fertile soil. According to Jagadeesha [31], sewage sludge treatment results in maximum moisture retention and the most efficient use of water. Studies show that using organic manure can improve water retention and effective rainfall, both of which serve to increase production when there is a water shortage.

Weed species	States in which it was reported as a major weed
Cyperus rotundus, Cynodon dactylon, Commelina benghalensis, Ageratum conyzoides, Echinochloa colona, Dactyloctenium aegyptium, Digitaria marginata, Eleusine indica, Spilanthes acmella, Acanthospermum hispidum, Eragrostis pilosa, Celosia argentea, Parthenium hysterophorus, Amaranthus viridis, Euphorbia hirta, Alternanthera sessilis, Digitaria sanguinalis (L.) Scop., *Leucas aspera, Sida accuta* Burm. f.	Andhra Pradesh, Bihar, Chhattisgarh, Gujarat, Orissa, Karnataka, Tamil Nadu, West Bengal, Uttar Pradesh,

Table 3.
Major weeds associated with finger millet in India [32].

Sr. N.	Management	Location state	References
1.	The conventional tillage (plowing twice + harrowing once + inter-cultivation twice at 25 and 50 days after sowing (DAS) in Alfisols when compared to minimum and zero tillage practices	Bangalore, Karnataka	[33]
2.	Hoeing twice by wheel hoe between rows + intra-row manual weeding *fb* HW twice 20 and 40 DAS	Raipur, Chhattisgarh	[34]
	2, 4-D sodium salt 0.75 kg/ha post-emergence application (PoE) 15–20 DAS	Bangalore, Karnataka; Berhampur, Orissa	[35, 36]
3.	Bensulfuron-methyl (0.6% G) + pretilachlor (6.0% G) 0.75 kg/ha (ready-mix) pre-emergence application (PE) (3 DAS)	Bangalore, Karnataka	[37, 38]
4.	Oxyfluorfen 0.1 kg/ha PE *fb* HW once 20 DAP Oxyfluorfen 0.1 kg/haPE	Tirupati, Andhra Pradesh Mandya, Karnataka, India	[39, 40]
5.	Bensulfuron-methyl 60 g + pretilachlor 600 g (6.6% G pre-mix formulation) 1.0 kg/ha pre-emergence application (PE) 2 DAP	Mandya, Karnataka	[41]

Table 4.
Method of weed control.

8. Weed management

Finger millet adapts well in adverse environmental conditions. Weed associated with finger millet is also adapted to those unfavourable conditions to compete with finger millet for the limited resources. Hence, it is essential to understand the ecology of weeds associated with finger millet to manage them properly (**Tables 3** and **4**) [42].

When finger millet is exposed to harsh environmental conditions, it adapts well and rapidly. In order to finger millet, compete with weeds for the resources and weed affects the overall development of the crop. Therefore, in order to effectively control of weeds associated with finger millet is crucial to understand their ecology [42]. The proper control of weeds increases finger millet yields. Integrated weed management (IWM) in the combination of herbicides, mechanical weeding methods, and hand weeding techniques successfully controlled the weeds in finger millet crop. IWM efficiently controls the losses caused by weeds like lowers nutrient uptake, which makes nutrients available to finger millet and lowers the cost of applying surplus fertilizers [43].

Sr. N.	Name of insect	Damaging symptoms	Management
1.	Cutworms (*Spodoptera exigua*)	The caterpillars cut seedlings at the base during early stage, which appears as if grazed by domestic animal. They are active during night and hide under stones and clods during the day.	After harvesting of crop remove weeds from the fields. Spray Quinalphos 20 EC @ 2 l/ha, Carbaryl 50 WP @ 2.5 kg/ha, or Phoshalone 35 EC @ 1.25 l/ha. Apply the biological agent's like-*entomopathogenic* nematode *Steinernema carpocapsae* or the fungus *Beauveria bassiana* when cutworms first emerge.
2.	Pink stem borer (*Sesamia inferens*)	Pink larva enters into the stem and causes dead heart symptoms.	Spray cartop hydrochloride 4G @ 25 kg/ha, fipronil 0.3G 15kg chlorpyriphos 10G 10kg in whorls.
3.	Aphids (*Rhopalosiphum maidis*	The Aphid Colonies present on the central leaf whorl and ears and after that yellowing of leaves. The appearance of ants at the infected plant parts.	Spray any one of the following insecticides mixed in 10 l of water using a high-volume sprayer if dusting is not done: Methyl demeton 25 4EC 20 ml/ha Dimethoate 30 EC 20 ml/ha
4.	White Grub, *Holotrichia consanguinea*	Grubs feed on roots which results in the death of the grown-up plants.	After the summer rains, initiate a widespread operation to collect and eliminate adult beetles that are hiding out in in the field. Use *Beauveria brongniortii* @ 2.5 kg/ha (1*109 cfu/g) *entomopathogenic* fungal formulation combined with FYM.
5.	Earhead caterpillars (*Sitotroga cerealella*)	Earhead caterpillars appear at the dough stage on ears and persist till harvest. The caterpillars bite the maturing seeds and make a fine web out of their casting and half-eaten grains.	Dust *Malathion* 5% @ 24 kg/ha or *Quinolphos* 1.5% @ 24 kg/ha.

Table 5.
Insects and their management [44, 45].

Sr. N.	Name of diseases	Damaging symptoms	Management
1.	Blast (*Pyricularia grisea*)	Young seedlings are infected in the nursery and in the fields, with spindle-shaped lesions of various sizes. Typical blast lesions are diamond shaped with a gray center and dark margin appearing on the leaf. The lesions are like those on the seedlings and are about 0.3–1.0 cm in breadth and about 1–2 cm in length.	The cultivation of resistant varieties can manage it. A day before planting, treat the seeds with fungicides such carbendazim @ 2 g/kg seed. Spray Tricyclazole (0.1%) or Carbendazim (0.1%) in the nursery if necessary. Use any of the fungicides suggested above at the flowering stage and repeat 10 days later.
2.	Leaf spot (*Helminthosporium nodulosum*)	Progressive infection from older to younger leaves can be observed in the standing crop. The leaf, leaf sheath, other plant parts are affected and shows small to medium-sized brown to dark spots.	The disease can be effectively managed by proper nutrition and water management. Need-based spraying of *Mancozeb* (0.2%) can be applied. However, field sanitation and spraying of Carbendazim at the rate of 0.05% has been reported to reduce infection to some extent.
3.	Downy Mildew or Green Ear Disease (*Sclerophthora macrospora*)	The plant assumes a bunchy and bushy appearance. Often, pale yellow translucent spots are seen on the leaves of affected plants. The white cottony growth, characteristics of many downy mildews, is generally not seen in the downy mildew of finger millet.	Affected plants should be destroyed by burning. Keep the field clean. In case of severe attack spray Mancozeb on the standing crop at the rate of 2 g/l of water.
4.	Bacterial Leaf Spot (*Xanthomonas campestris pv. Eleusineae*)	The typical symptom appears as light yellowish to brown dots on the leaf's both surfaces. They extend along the veins in a linear pattern. Subsequently, the color turns dark brown, and as the illness worsens, the leaf blade splits along the streaks.	The spray of mancozeb @ 2 g/l or copper oxychloride @ 2.5 g/l. Seed treatment with 0.1% mercuric chloride solution for 2–5 minutes is effective. Spraying with streptomycin should not be done after fruits begin to form. Field sanitation is important. Also, seeds must be obtained from disease-free plants.

Table 6.
Diseases and their management.

9. Plant protection

9.1 Insect pests and their management

Finger millet attracts several pests of which armyworms, cutworms, stem borers, leaf aphids, grasshoppers, gray weevils, shoot flies, and ear caterpillars are major ones (**Tables 5** and **6**).

10. Harvesting and post-harvest management

When millet grains reach full physical maturity, they should be harvested. Due to quick temperature and humidity changes, late harvesting could result in losses and grain quality deterioration. Millet can be threshed manually or mechanically, including by pounding the panicles with sticks. When the crop is harvested by hand, the panicle is separated from the upright stalk at a moisture level of approximately 16–20%, and the stalks are used as animal feed. Before threshing, the grain's dryness can be determined by biting into it with the teeth or pinching it between the fingers. A dry glass bottle is filled with dry salt, the grain is added, and the bottle is shaken to test the salt. If, after a few minutes, the salt adheres to the bottle's sides, the grain's moisture content is above 15%; if not, the grain's moisture content is correct. Using the teeth (the grains are brittle when bitten) or pinching with the fingers, one can determine the dryness of the grain prior to threshing [29]. The grain must be dried once more to a moisture content of 13% or less after it has been threshed. The threshed grain should be dried on wire mesh trays, mats, sheets, or tarpaulins made of plastic that have been elevated on a platform. To allow air to travel through the grain while it dries, spread it out thin on the drying surface. Be sure to move the grain frequently to prevent overheating. Keep the grain dry and free from dirt, insects, animals, and rain [46, 47].

11. Utilization of finger millet

Even though finger millets have a wide variety and excellent nutritional content. The start-up movement of finger millet to increase the availability of nutrient-rich food has recently been steadily fuelled by these grains [48]. Tribal people save 75% of their finger millet harvest for food, therefore eating it every day is a ritual. The availability of finger millet is found to be practically year-round in their homes 65% of households, but we still need to address the remaining 35% of families to ensure year-round availability. Most families believe they eat finger millet because it provides more energy than other foods and makes it easy to carry out daily tasks, and they are least likely to be aware of blood sugar regulation [49]. Processing may be two different types: main processing and secondary processing. Primary processing consists of operations like cleaning, washing (soaking/germination), dehulling, milling (into flour and semolina), and removing the undesirable seed coat and anti-nutritional factors, while secondary processing entails transforming raw materials into "ready-to-cook" (RTC) or "ready-to-eat" (RTE) products through flaking, popping, extrusion, and baking [50]. In India, malting finger millet is a common practice that produces ragi malt and a milk thickening practice that is used in baby food and other products. The preparation of products made only of millet and the mixing of millet with other ingredients can be done using a variety of techniques. These techniques can be similar to those used to prepare products from wheat and rice or they can differ because finger millets have different physical-chemical characteristics than the other cereal grains. The grain is also malted, and the flour produced from the malted grain is used to feed newborns and the elderly [51]. The development of a pleasant aroma during the kilning of the germinated grain is an additional benefit of malting ragi. In order to make milk drinks, malted ragi flour, also known as "ragi malt," is utilized. In certain regions of the nation, a fermented beverage called beer is also made from grain [7, 52].

12. Conclusion

This crop is perfect for dry land farming because of its strong ability for regeneration after the relief of stressful circumstances. The major states in India where finger millet is grown include Karnataka, Uttarakhand, Tamil Nadu, Andhra Pradesh, Orissa, Jharkhand, and Maharashtra. Furthermore, finger millet is a crucial component of dietary and nutritionally balanced diets since it is equally rich in carbohydrates, energy, and nutrients. Despite the fact that finger millet is a very healthy grain with a high nutritional value, a high consumption may cause the body to produce more oxalic acid. Compared to other cereals, millets are nutritionally dense, and processing and employing millets in the production of products has undeniable potential in terms of health benefits, nutrition, and quality. By increasing the digestibility of the protein and the bioavailability of the minerals, simple processing methods like soaking, germination/malting, and fermentation may assist address the issue of protein-energy deficiency. In order to increase consumer acceptance of small millets without sacrificing their health advantages, processing procedures for these grains still need to be optimized. Additionally, knowledge must be raised about the effects of processing techniques on millets' nutritional qualities and health advantages at both the commercial and household levels in order to address food poverty and malnutrition.

Author details

Brijesh Kumar[1*], Magan Singh[1], Avaneesh Kumar[1], Deepak Kumar[1], Ankit Kumar Singh[2], Rakesh Kumar[1] and Sandeep Kumar[1]

1 Agronomy Section, ICAR-National Dairy Research Institute, Karnal, Haryana, India

2 Division of Plant Pathology, Chaudhry Charan Singh Haryana Agriculture University, Hissar, Haryana, India

*Address all correspondence to: kbrijeshkumar609@gmail.com

References

[1] Chandra D, Chandra S, Sharma AK. Review of Finger millet (*Eleusine coracana* (L.) Gaertn): A powerhouse of health benefiting nutrients. Food Science and Human Wellness. 2016;**5**:149-155

[2] Upadhyaya HD, Gowda CL, Reddy VG. Morphological diversity in finger millet germplasm introduced from Southern and Eastern Africa. Journal of SAT Agricultural Research. 2007;**3**(1):1-3

[3] Gull A, Jan R, Nayik GA, Prasad K, Kumar P. Significance of finger millet in nutrition, health and value-added products: A review. Magnesium (mg). 2014;**130**(32):120

[4] Antony Ceasar S, Maharajan T, Ajeesh Krishna TP, Ramakrishnan M, Victor Roch G, Satish L, et al. Finger millet [*Eleusine coracana* (L.) Gaertn.] improvement: Current status and future interventions of whole genome sequence. Frontiers in Plant Science. 2018;**9**:1054

[5] Krishna TA, Maharajan T, Roch GV, Ramakrishnan M, Ceasar SA, Ignacimuthu S. Hybridization and hybrid detection through molecular markers in finger millet [*Eleusine coracana* (L.) Gaertn.]. Journal of Crop Improvement. 2020;**34**(3):335-355

[6] Maharajan T, Ceasar SA, Krishna TP, Ignacimuthu S. Phosphate supply influenced the growth, yield and expression of PHT1 family phosphate transporters in seven millets. Planta. 2019;**250**(5):1433-1448

[7] Saleh AS, Zhang Q, Chen J, Shen Q. Millet grains: Nutritional quality, processing, and potential health benefits. Comprehensive Reviews in Food Science and Food Safety. 2013;**12**:281-295

[8] Banerjee P, Maitra S, Banerjee P. The role of small millet as functional food to combat malnutrition in developing countries. Indian Journal of Natural Sciences. 2020;**10**(60):20412-20417

[9] Chandrasekara A, Shahidi F. Content of insoluble bound phenolics in millets and their contribution to antioxidant capacity. Journal of Agricultural and Food Chemistry. 2010;**58**(11):6706-6714

[10] Singh RB, Khan S, Chauhan AK, Singh M, Jaglan P, Yadav P, et al. Millets as functional food, a gift from Asia to Western world. In: The role of functional food security in Global Health. Academic Press. 2019. pp. 457-468. DOI: 10.1016/B978-0-12-813148-0.00027-X

[11] Parameswaran KP, Sadasivam S. Changes in the carbohydrates and nitrogenous components during germination of proso millet, *Panicum miliaceum*. Plant Foods for Human Nutrition. 1994;**45**(2):97-102

[12] Antony Ceasar S, Ignacimuthu S. Agrobacterium-mediated transformation of finger millet (*Eleusine coracana* (L.) Gaertn) using shoot apex explants. Plant Cell Reports. 2011;**30**(9):1759-1770

[13] Ravindran G. Studies on millets: Proximate composition, mineral composition, and phytate and oxalate contents. Food Chemistry. 1991;**39**:99-107

[14] Sripriya G, Antony U, Chandra TS. Changes in carbohydrate, free amino acids, organic acids, phytate and HCl extractability of minerals during germination and fermentation of finger millet (*Eleusine coracana*). Food Chemistry. 1997;**58**(4):345-350

[15] DeviPB,VijayabharathiR,SathyabamaS, Malleshi NG, Priyadarisini VB. Health benefits of finger millet (*Eleusine coracana* L.) polyphenols and dietary fiber: A review. Journal Food Science Technology. 2014;**51**:1021-1040

[16] Scalbert A, Manach C, Morand C, Rémésy C, Jiménez L. Dietary polyphenols and the prevention of diseases. Critical Reviews in Food Science and Nutrition. 2005;**45**(4):287-306

[17] McKeown NM, Meigs JB, Liu S, Wilson PW, Jacques PF. Whole-grain intake is favourably associated with metabolic risk factors for type 2 diabetes and cardiovascular disease in the Framingham offspring study. The American Journal of Clinical Nutrition. 2002;**76**(2):390-398

[18] Available from: https://eands.dacnet.nic.in/Advance_Estimate/Time%20Series%204%20AE%202021-22%20(English).pdf

[19] Sakamma S, Umesh KB, Girish MR, Ravi SC, Satishkumar M, Bellundagi V. Finger millet (*Eleusine coracana* L. Gaertn.) production system: Status, potential, constraints and implications for improving small farmer's welfare. The Journal of Agricultural Science. 2018;**10**(1):162-179

[20] Chapke RR, Shyam Prasad G, Das IK, Hariprasanna K, Singode A, Kanthi Sri BS and Tonapi VA. 2020. Latest Millet Production and Processing Technologies. Booklet, ICAR-Indian Institute of Millets Research, Hyderabad 500 030, India: 82p. (ISBN: 81-89335-90-X)

[21] Sukanya TS, Prabhakar KG, KT, Swarna R, Hariprasanna K and Tonapi VA. Good agronomic practices for higher yield in small millets. In: ICA—All India Coordinated Research Project on Small Millets. Hyderabad: ICAR-Indian Institute of Millets Research; 2022

[22] Rao KL, Rao CP, Rao KV. Response of finger miller (eleusine-coracana l gaertin) cultivars to nitrogen under rain-fed conditions. Indian Journal of Agronomy. 1989;**34**(3):302-306

[23] Hegde BR, Gowda L. Cropping systems and production technology for small millets in India. In: Proceedings of the First International Small Millets Workshop, Bangalore, India, New Delhi: Oxford and IBH publishing. 1989. pp. 209-236

[24] Srinivasa Rao C, Wani SP, Sahrawat KL, Rego TJ, Pardhasaradhi G. Zinc, boron and Sulphur deficiencies are holding back the potential of rainfed crops in semi-arid India: Experiences from participatory watershed management. International Journal of Plant Production. 2008;**2**(1):89-99

[25] Rajashekhara Rao BK, Krishnappa K, Srinivasarao C, Wani SP, Sahrawat KL, Pardhasaradhi G. Alleviation of multi-nutrient deficiency for productivity enhancement of rain-fed soybean and finger millet in the semi-arid region of India. Communications in Soil Science and Plant Analysis. 2012;**43**(10):1427-1435

[26] Ramakrishnan K, Bhuvaneswari G. Effect of inoculation of am fungi and beneficial microorganisms on growth and nutrient uptake of *Eleusine coracana* (L.) Gaertn.(Finger millet). International Letters of Natural Sciences. 2014;**8**(2):81-91

[27] Tewari L, Johri BN, Tandon SM. Host genotype dependency and growth-enhancing ability of VA-mycorrhizal fungi for Eleusine coracana (finger millet). World Journal of Microbiology and Biotechnology. 1993;**9**(2):191-195

[28] Apoorva KB, Prakash SS, Rajesh NL, Nandini B. STCR approach for optimizing integrated plant nutrient supply on growth, yield and economics of finger millet (*Eleusine coracana* (L.) Garten.). EJBS. 2010;**4**(1):19-27

[29] Meena DS, Chirag G, Patidar O, Ranvir S, Meena H, Vishwajit PG. Management of finger millet-based cropping systems for sustainable production. International Journal of Current Microbiology and Applied Sciences. 2017;**6**(3):676-686

[30] Jagadeesha N. Productivity of Finger millet (Eleusine coracana(L.) Gaertn.) and Redgram (Cajanuscajan (L.) Millsp.) Intercropping System as Influenced by Organic Sources. Ph.D. thesis,. Bengaluru, Karnataka: UAS; 2009

[31] Gupta SM, Arora S, Mirza N, Pande A, Lata C, Puranik S, et al. Finger millet: A "certain" crop for an "uncertain" future and a solution to food insecurity and hidden hunger under stressful environments. Frontiers in Plant Science. 2017;**8**:643

[32] Hatti V, Ramachandrappa BK. Weed Dynamics in Conservation Agricultural Systems as Influenced by Conservation Tillage and Nutrient Management Practices under Rainfed Finger millet. Indian Journal of Weed Science. 2018;**50**(4):355-364

[33] Kujur S, Singh VK, Gupta DK, Tandon A, Ekka V, Agrawal HP. Influence of weed management practices on weeds, yield and economics of finger millet (*Eleusine coracana* L. Gaertn). International Journal of Bioresource and Stress Management. 2018;**9**:209-213

[34] Jena BK, Tripathy SK. Effect of weed control in finger millet (*Eleusine coracana*). Indian Journal of Agronomy. 1997;**42**:641-644

[35] Ashok EG, Chandrappa M, Kadalli GG, Kiran K, Mathad V, Gowda KT. Integrated weed control in drill-sown rainfed finger millet (*Eleusine coracana*). Indian Journal of Agronomy. 2003;**48**:290-293

[36] Kumar MKP, Shekara BG, Sunil CM, Yamuna BG. Response of drill sown finger millet to pre and post-emergent herbicides. The Bioscan. 2015;**10**:299-302

[37] Kumar MKP, Shekara BG, Yamuna BG, Sunil CM. Crop weed competition for nutrients by weeds and drill sown finger millet (*Eleusine coracana* (L.) Gaertn.). International Journal of Tropical Agriculture. 2015a;**33**:2049-2053

[38] Bhargavi B, Sunitha N, Reddy YR, Reddy GP. Efficacy of herbicides on weed suppression in transplanted finger millet (*Eleusine coracana*). Indian Journal of Agronomy. 2016;**61**(1):109-111

[39] Prakash P, Ravishankar R, Ramappa HK, Geethadevi T. Millets Research and Development – Future Policy Options in India. 2006. pp. 38-39

[40] Gowda SGK, Naveen DV, Bhagyalakshmi T, Gowda RC. Weed management practices on nutrient removal by weeds and its relation to yield of finger millet in eastern dry zone of Karnataka. International Journal of Agriculture Sciences. 2012;**8**:385-389

[41] Banu A, Fathima PS, Denesh GR, Sunil CM. Pre and post-emergence herbicides for weed management in finger millet. Indian Journal of Weed Science. 2016;**48**:447-449

[42] Rao AN. Weed management in finger millet in India-an overview. Indian Journal of Weed Science. 2021;**53**(4):324-335

[43] Available from: http://eagri.org/eagri50/ENTO331/lecture03/finger_millet/index.html

[44] Available from: https://agritech.tnau.ac.in/crop_protection/crop_prot_crop_insectpest%20_cereals_ragimain.htm

[45] Available from: https://agritech.tnau.ac.in/crop_protection/crop_prot_crop%20diseases_cereals_ragi_main.html

[46] Available from: https://www.millets.res.in/books/DISEASES_OF_MILLETS.pdf

[47] https://agritech.tnau.ac.in/postharvest/pht_millets_ragi.html

[48] Annor GA, Tyl C, Marcone M, Ragaee S, Marti A. Why do millets have slower starch and protein digestibility than other cereals? Trends in Food Science & Technology. 2017 Aug;**1**(66):73-83

[49] https://tcrconnectingagriculture.com/2020/07/study-on-utilisation-of-finger-millet-in-nutritional-security/

[50] Dayakar Rao B, Bhaskarachary K, Arlene Christina GD, Sudha Devi G, Vilas AT, Tonapi A. Nutritional and health benefits of millets. Hyderabad, Indian: ICAR_Indian Institute of Millets Research (IIMR); 2017. p. 112

[51] Birania S, Rohilla P, Kumar R, Kumar N. Postharvest processing of millets: A review on value-added products. International Journal of Chemical Studies. 2020;**8**(1):1824-1829

[52] Nithiyanantham S, Kalaiselvi P, Mahomoodally MF, Zengin G, Abirami A, Srinivasan G. Nutritional and functional roles of millets—A review. Journal of Food Biochemistry. 2019;**43**(7) e12859

Chapter 4

Biofortification of Millets: A Way to Ensure Nutritional Security

R.K. Anushree, Shailja Durgapal, Meenal and Latika Yadav

Abstract

Malnutrition poses significant socioeconomic challenges worldwide, with its most acute impact felt in developing and impoverished nations. This issue is exacerbated by the reliance on cereal-based diets, which often lack essential micronutrients, as the world's population continues to grow. Millets and whole grains emerge as promising solutions to this dilemma. Although millets have historically served as a primary energy source in regions like Asia, Africa, and other semi-arid tropical areas, their nutritional value has been underrated. Beyond their carbohydrate content, millets are rich sources of vitamins, minerals, and crucial amino acids. Biofortification, the practice of enhancing the nutrient content of staple crops, offers a cost-effective approach to address micronutrient deficiencies. Initiatives like Harvest Plus in India have introduced biofortified millets to combat widespread deficiency disorders. The global distribution of biofortified millets is supported by non-governmental organisations, the business sector, and government regulatory bodies. This book chapter delves into these critical efforts, emphasising their role in ensuring nutritional security and effectively tackling malnutrition on a global scale.

Keywords: malnutrition, nutritional security, harvest plus, biofortification, micronutrient deficiencies

1. Introduction

Malnutrition has substantial socioeconomic repercussions everywhere, but is more prevalent in developing and underdeveloped countries. Malnutrition, which is caused by insufficient intake of a balanced diet, compromises health, increases vulnerability to numerous diseases, and causes a large loss in annual Gross Domestic Product (GDP), which can reach 11% in Asia and Africa [1]. Around 2 billion people around the world are undernourished, and 815 million are malnourished. Malnutrition has the greatest impact on children, who account for 151 million stunted children under the age of 5 and 51 million who do not weigh enough for their height (wasting) Malnutrition is linked to over 45% of fatalities in children under the age of five [2]. The issue is so pervasive that hunger in two or more forms affects 88% of the world's nations. Malnutrition is most prevalent in Southern Asia, where 33.3 and 15.3 percent of children (>5 years) are stunted and wasted, respectively, compared to the global averages of 22.2 and 7.5 percent [3].

The process of "biofortification" involves using standard plant breeding or agronomic techniques, like fertiliser administration, to increase the amount of micronutrients in food crops during plant growth [4]. It is an agricultural-nutrition strategy that addresses the most prevalent and avoidable global micronutrient gaps in populations that depend on staple food crops for sustenance and have little access to alternative sources of micronutrients, such as fortified foods, supplements, or more varied diets, and have physiological needs that differ from intake [5, 6].

Millets, which produce an average of 14.2 and 12.4 million tonnes annually and are consumed in resource-poor nations in Asia and Africa, make about 75% of the total calories consumed. About 80% of the world's millet production is produced in India, making it the top producer of millets [7]. The term "small seeded grasses" is often used to describe millets, which include pearl millet (Pennisetum glaucum (L.) R. Br.), finger millet (Eleusine coracana (L.) Gaertn), foxtail millet (Setariaitalica (L.) Beauv), proso millet (Panicum miliaceum L.), barnyard millet (Echinochloa (Panicum sumatrense) [8–11].

Pearl millet accounts for 95% of the production of the millets, The second-largest crop of millets, foxtail millet (*S. italica* (L.) P. Beauv) is grown for food in Asia's semi-arid tropics and for forage in Europe, North America, Australia, and North Africa [12]. The sixth-largest crop now grown, finger millet is the main source of nutrition for rural inhabitants in southern India, East and Central Africa. A short-season crop called proso millet is grown in arid areas of Asia, Africa, Europe, Australia, and North America [13, 14]. Barnyard millet is the fastest growing among the millets with a harvesting period of 6 weeks [15]. It is mostly grown for food and fodder in India, China, Japan, and Korea. Native to South America's tropical and subtropical climates, kodo millet was domesticated in India. Before 3000 years [16, 17].

Because millets are rich in proteins, dietary fibres, iron, zinc, calcium, phosphorus, potassium, vitamin B, and vital amino acids, they are nutritionally superior to wheat and rice [18, 19]. Phytates, polyphenols, and tannins, however, are antinutrients that chelate multivalent cations including Fe2+, Zn2+, Ca2+, Mg2+, and K+ to lower the bioavailability of minerals [20–24]. Additionally, the digestibility of millet grains is impacted by high levels of protease and amylase inhibitors [25–27]. Millets are now considered to be an economic outcast on the global stage due to the domination of antinutritional elements. With the exception of Golden rice, biofortified crops have generally been generated through conventional breeding that takes use of the genetic diversity present in the environment (www.harvestplus.org). When compared to other cereal crops, millets have significantly higher genetic variability for important mineral elements including iron, zinc, and calcium [28]. Moreover, millets are drought tolerant crops [29], resistant to pests and diseases offering good insurance against crop failure in developing countries [30, 31]. Despite millets' higher quality, India has solely prioritised pearl millet as the preferred crop for iron biofortification. As a result, there is enormous potential to use the minor millets for biofortification. Millets can be biofortified using one of two methods: either increasing the nutrient accumulation in milled grains or lowering the antinutrients to boost the bioavailability of minerals. This book chapter will emphasise in-depth details on millets' biofortification.

2. Methodology for the review of the literature

PubMed, Google, and other databases are searched for relevant material. We conducted a search of all review papers using the keywords "Millets, Bio fortification,

Nutritional Security." Additionally, the global scenario, efforts, critical evaluations, government reports, agency reports, and publicly available data were analysed. The necessary data was gathered, compiled, and analysed.

3. Nutrients and their functions

3.1 Nutritional factors

a. *Protein:* It provides the necessary amino acids for tissue repair and growth. Poor intellectual growth, disorganised physical functioning, and even fatality result from its lack. Humans that consume inadequate amounts of protein develop the conditions kwashiorkor and marasmus [32].

b. *Lysine:* Along with acting as a precursor for a number of neurotransmitters and metabolic regulators, it is a component of protein synthesis. Lysine deficiency causes exhaustion, nausea, dizziness, anaemia, slowed growth, appetite loss, and deterioration of reproductive tissue.

c. *Tryptophan*: It serves as a precursor for a number of neurotransmitters and as a regulator of metabolic pathways in addition to being a protein building block. Lack of it causes irritability, anxiety, and depression. The main signs of tryptophan insufficiency in children include weight loss and delayed growth.

d. *Iron:* The healthy functioning of the brain and muscular tissues depends on this mineral element. Oxygen is transferred from the lungs to other tissues via red blood cell haemoglobin. The most typical sign of iron deficiency in humans is the development of anaemia. Growth and development are also delayed by iron deficiency.

e. *Zinc*: It is a mineral substance that serves as a cofactor in more than 300 vital human enzymes. Controlling the production and breakdown of proteins, lipids, carbohydrates, and nucleic acids depends on it. Zinc deficiency causes growth slowdown, appetite loss, compromised immune system function, and increased susceptibility to infections.

f. *Calcium:* It is a mineral element necessary for the development and maintenance of healthy bones and teeth. Additionally, it affects how your heart and muscles move. Osteoporosis, which results from a calcium deficiency, makes bones fragile. The other related symptoms include dental issues, cataracts, and changes in the brain.

g. *Vitamin-A:* It is essential for the immune system, reproduction, growth, and development, the maintenance of epithelial cell integrity, and the regular operation of the visual system. It is also referred to as "retinol." The defining symptom of vitamin A deficiency is night blindness. Conjunctival and corneal structural changes may also result in xerophthalmia and keratomalacia. Additionally increased risks include those for anaemia, diarrhoea, measles, malaria, and respiratory infections.

h. *Vitamin-C:* Many tissues, including cartilage, bone, skin, and teeth, require it for metabolism and repair. The digestive system's capacity to absorb iron is improved by vitamin C. Gum bleeding, bruising, and a poor ability to heal tooth wounds are symptoms of scurvy, which is brought on by a deficiency. Muscle and joint issues are also connected to it.

i. *Anthocyanins:* These pigments give plant parts their red, purple, and blue hues. Anthocyanins function as antioxidants and aid in the removal of dangerous free radicals generated within the body. In addition to preventing cardiovascular illnesses, anthocyanins have anti-obesity, anti-inflammatory, anti-cancer, anti-inflammatory, anti-microbial, and anti-cancer properties.

j. *Oleic acid:* It is a mono unsaturated fatty acid present in oil. Low-density lipo-protein (LDL) cholesterol levels and the risk of coronary heart disease are both associated with dietary monounsaturated fat intake.

k. *Linoleic acid:* It is a type of oil-found polyunsaturated fatty acid. Because it lowers total and LDL cholesterol, it's beneficial for cardiovascular health.

3.2 Anti-nutritional factors

a. *Erucic acid:* Mustard,rapeseed oils etc. contain this monounsaturated fatty acid. High erucic acid content in edible oils raises blood cholesterol in children, decreases cardiac conductivity, and results in lipidosis [32].

b. *Glucosinolates*: This group of thioglucosides is mostly found in the Brassicaceae family. When myrosinase breaks down glycosinolates, additional compounds, including glucose and sulphate, are produced. Excessive intake is detrimental to the health of animals since it reduces the appeal of feed and obstructs the thyroid gland's capacity to absorb iodine, which decreases feed efficiency and weight growth, especially in non-ruminants like pigs and poultry.

c. *Kunitz trypsin inhibitor (KTI):* It is a non-glycosylated protein that stunts human growth, especially by preventing the digestive enzyme trypsin from working and producing dyspepsia. Most trypsin inhibitors in soybeans, including KTI, are thought to be hazardous to human health.

d. *Lipoxygenase:* It is an enzyme that aids in the oxidation of polyunsaturated fatty acids, which imparts an unpleasant flavour to foods made from soybeans. The off-flavour of beans makes customers less likely to prefer soybean as food.

4. Sustainable development goals (SDGs)

The United Nations (UN) developed 17 Sustainable Development Goals (SDGs) in 2015 to chart a route for meeting present-day human needs without compromising the ability of future generations to achieve the same. The core goals of the SDGs are to safeguard the environment, ensure that everyone lives in peace and prosperity, and eradicate extreme poverty, hunger, and malnutrition by 2030. 12 of the 17 goal-indicators have a connection to nutrition [33].

5. Status of malnutrition: Indian scenario

- 35.7% of kids are underweight, 21.0% are wasted, and 38.4% are stunted before the age of five. The population's extreme poverty rate is 21.9% [33].

- Stunting rates vary greatly among districts (12.4–65.1%), and out of 640 districts, 239 have stunting levels above 40% [34].

- 53% of adult women, 22.7% of adult males, and 58.4% of newborns (6–59 months) are affected by anaemia [35].

- According to estimates, 38% of children (under the age of 5) are zinc deficient, and 70% of children (under the age of 5) are iron deficient, which costs India's GDP roughly USD 12 billion yearly [36].

By the creation of high-yielding biofortified crop varieties, ICAR is dedicated to the SDGs by taking into account all of these factors.

6. Biofortified crop cultivars

The National Agricultural Research System (NARS), which consists of ICAR institutes and State Agricultural Universities, has significantly contributed to India's achievement of food self-sufficiency (SAUs). The amount of food produced in India increased from 50.82 mt in 1950–1951 to 284.8 mt in 2017–2018. (Fourth Advance Estimates) 24. Horticulture crops also increased, going from 96.56 mt in 1991–1992 to 306.8 mt in 2017–2018. (Third Advance Estimates) 24. The tremendous increase in yield potential has been made possible through the development and application of high yielding cultivars and heterotic hybrids, activities that began during the Green Revolution. As of this writing, NARS has produced 4723 unique field crop varieties. Yet during the process of yield augmentation, nutritional quality was not given the attention it required, and as a result, the bulk of these varieties lack the proper level of nutritional quality. Recognising the crucial importance of nutritional quality, NARS' research has now produced and distributed a variety of biofortified varieties for different crops through All Indian Coordinated Research Programmes (AICRPs). The biofortified varieties also offer enough calories while also supplying the requisite nutrient(s) for healthy growth and development [7].

Similar advantages More than 2 billion people, or one in three people, experience vitamin deficiencies globally. Such deficiencies occur when dietary intake and mineral absorption are insufficient to sustain healthy growth and development. Agricultural research for developing countries has boosted the production and accessibility of calorically dense staple crops during the past 50 years, but not in a proportionate way for non-staples like vegetables, pulses, and animal products, which are high in micronutrients. It has become increasingly difficult for the poor to afford dietary quality as the price of non-essential goods has constantly and considerably grown [7].

Through increasing production of foods high in micronutrients and diversifying diets, long-term reductions in micronutrient shortages will be possible. Eating crops that have been biofortified can help reduce short-term micronutrient shortages by increasing daily adequacy of micronutrient intakes across persons throughout their lives. Biofortification is a helpful complement to other therapies, such as dietary supplements and commercial food fortification, for treating micronutrient deficits that cannot be remedied with a single action. To reach underserved rural populations and to be long-term cost-effective, biofortification offers two important comparative advantages. Contrary to the ongoing financial commitments required for supplementation and commercial fortification projects, a one-time investment in plant breeding yields micronutrient-rich biofortified planting material for farmers to grow at nearly zero marginal cost [37].

After being grown, crops with improved nutrition can be evaluated and adjusted to different environments and areas, double the benefits of the initial investment. When the micronutrient trait is incorporated into the fundamental breeding objectives of national and international crop development programmes, ongoing expenses for monitoring and maintenance by agriculture research institutes are minimal. Another practical aim of biofortified crops is to reach rural populations that might struggle to acquire healthy food or other micronutrient treatments. The target micronutrient levels for biofortified crops are determined to meet the specific nutritional needs of women and children based on current consumption trends. Farmers now have a choice thanks to biofortification, which combines the micronutrient trait with other desired agronomic and consumer qualities [37].

6.1 Crop development

Plant breeding can increase staple crop nutrient levels to target levels needed for increasing human nutrition without losing yield or farmer-preferred agronomic traits. It is required to screen germplasm for genetic diversity, design and test germplasm that is rich in micronutrients, conduct genetic research, and develop molecular markers to expedite and lower the cost of breeding in order to create novel crops. After being created, promising lines are evaluated in various locations throughout target habitats to determine the genotype x environment interaction (GxE), or the effect of the growing environment on micronutrient expression. Strong regional testing enables the reduction of time to market for biofortified cultivars Nutritional breeding targets for each crop were established early on in the conceptual development of biofortification by a working group comprising nutritionists, food technologists, and plant breeders. These goals were established based on the food consumption habits of the target groups, expected nutrient losses during storage and processing, and nutrient bioavailability [38].

When developing breeding targets for biofortified crops, consideration was given to the particular dietary needs and eating patterns of women and children. Targets were set so that, for preschoolers ages 4–6 and for non-pregnant, non-lactating women of reproductive age, the total amount of iron in iron beans and iron pearl millet will provide roughly 60% of the Estimated Average Requirement (EAR) (30% of the EAR for iron at baseline before breeding for high iron); the amount of zinc in zinc wheat and zinc rice will provide 60–80% of the EAR (40% of the EA); and the total amount of zinc in (zero provitamin A at baseline). The breeding aim is the total of the baseline micronutrient content and the required increase in micronutrient content for each crop and micronutrient combination [38].

6.2 A framework for the breeding of bio-fortified germplasm

The major procedures for producing biofortified germplasm are depicted in **Figure 1**. To ensure nutrient impact and farmer and consumer consent, activities

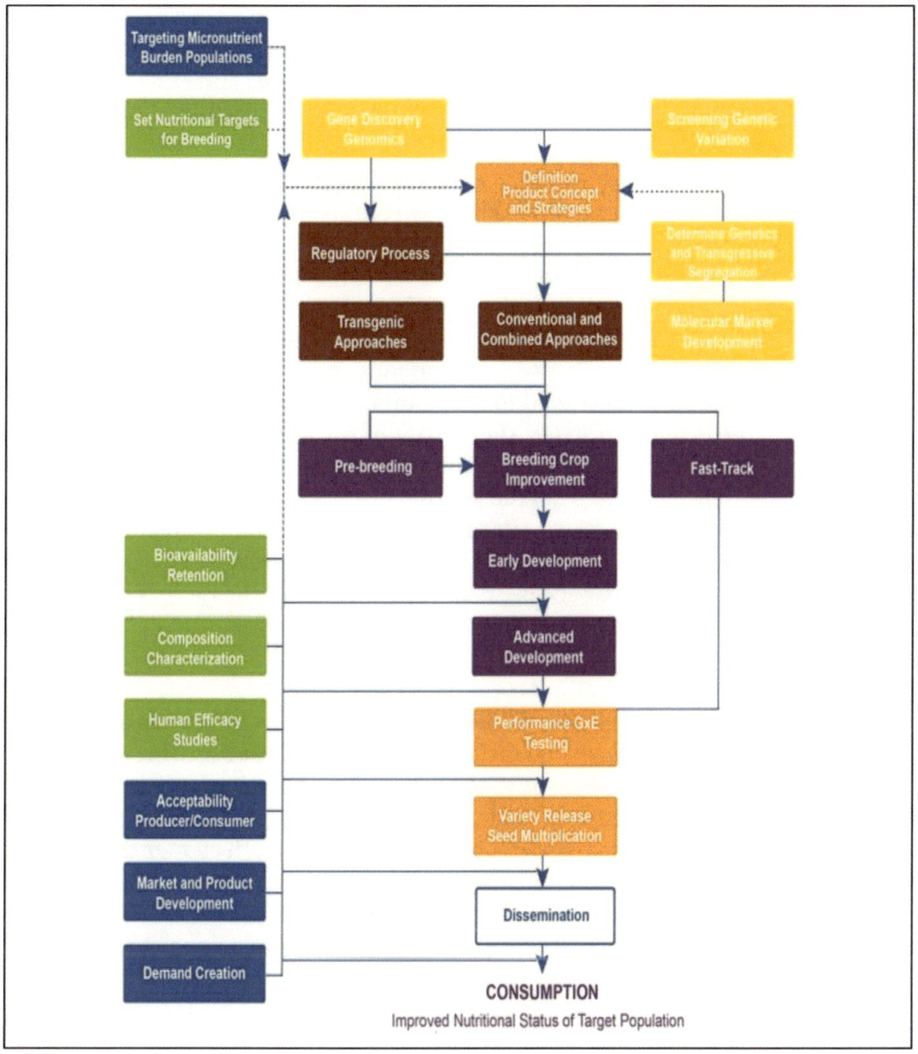

Figure 1.
Crop development framework [39].

outside of crop development are indicated in the left column [39]. A decision-tree that allows for tracking progress and making strategic decisions when goals are not fulfilled is placed above the right columns, which present the stages and milestones of crop development in sequential order. To design crops for biofortification, the first step is to look into the genetic variety that is now available for iron, zinc, and provitamin A carotenoids (yellow boxes). Characterisation of agronomic and end-use characteristics occurs immediately with or during subsequent screening.

When investigating the genetic diversity that exists, the following goals need to be noted:

1. Parental genotypes for cross-breeding, genetic research, creating molecular markers, and parent-building.

2. For "fast-tracking," choose pre-varieties that have already been released or finished germplasm products. Fast-tracking is the process of releasing, commercialising, or introducing genotypes that have the desired agronomic and end-use features as well as the target micronutrient density so they may be distributed right away.

If variation is present in the strategic gene pool (only in unadapted sources), pre-breeding is necessary before employing the trait in final product creation; if variation is present in the adapted gene pool, the materials can be used right away to create competitive variations (purple boxes). Prebreeding and product improvement activities are combined in the majority of breeding efforts to produce germplasm with high levels of one or more micronutrients. In the later stages of breeding, micronutrient-rich germplasm is developed and evaluated, genetic studies are conducted, and molecular markers are developed to speed up breeding. In trial locations and in farmer's fields in the target countries, the impact of the growing environment on the expression of micronutrients is then determined (orange boxes). The most promising varieties are chosen by national research partners for multi-locational testing over several seasons, and after being submitted to national government agencies for testing for agronomic performance and release, a procedure that typically takes 2 years, occasionally longer, are then tested for their performance in the field (**Figure 1**).

6.3 Transgenic approaches

When the desired nutrient does not naturally occur in the hundreds of varieties in germplasm banks at the necessary quantities, transgenic plant breeding is a promising way to generate biofortified crops with the requisite nutrient and agronomic properties. Restricted field tests, for instance, have been carried out on transgenic iron and zinc rice, which may provide 30% of the EAR for both elements [40]. As golden rice contains beta carotene, it can provide more than 50% of the EAR for vitamin A. Since early 2000, there has been a prototype of Golden Rice, but it has not yet been made available for purchase in any country, partly due to the regulatory clearance processes' extreme risk aversion [41]. Despite the fact that the introduction of these transgenic cultivars to farmers is still a few years away and is dependent on their approval through national biosafety and regulatory processes, they offer a large nutritional potential. For HarvestPlus efforts, all of the crops that have been produced or will be released soon employ traditional breeding rather than transgenic breeding. HarvestPlus thinks that because traditional breeding does not encounter the same regulatory hurdles and is widely accepted, it is the fastest way to get more nutrient-dense crops into the hands of farmers and consumers. The focus of this essay is the data offered in support of conventionally grown biofortified crops.

6.4 International nurseries/global testing

HarvestPlus has used two tactics to shorten the time to market for biofortified crops: Two techniques are being employed to quicken release operations while cultivars with the necessary micronutrient content are still being developed: (1) Choosing adapted varieties with high micronutrient contents for release and/or distribution as "quick track" varieties, and (2) conducting multi-location Regional Trials in numerous locations across a variety of countries and sites. Regional trials comprise biofortified varieties that have previously been released and generate data on their regional performance in order to benefit from regional variety release schemes, such as those under the SADC (Southern African Development Community). These regional agreements harmonise seed regulations among participants and enable the simultaneous distribution of any variety tried, approved, and released in one participant country in participants with comparable agro-ecologies [39].

6.5 Low-cost, high throughput methods

Biofortification breeding required the development or use of rapid, inexpensive analytical methods for micronutrients due to the necessity of analysing hundreds of samples for mineral or vitamin content each season. These trait diagnostics include methods like NIRS (near-infrared spectroscopy) and colorimetric carotenoid measurements. Since it involves minimal pre-analytical preparation and permits non-destructive inspection, X-ray fluorescence spectroscopy (XRF) has emerged as the method of choice for mineral analysis [42, 43].

6.6 Releases of biofortified crops

More than 150 biofortified cultivars of ten different crops have been sent to 30 different nations overall. A total of 12 different crops' potential biofortified types are being considered for distribution in 25 more countries. **Figure 2** depicts the areas where biofortified cultivars have been tested and made accessible thus far. Countries in the dark purple have already made biofortified crops available, while those in the light purple are still testing them. In the countries depicted on this map, the orange sweet potato has been propagated by the International Potato Center (CIP). You may get more particular information about the cultivars that have been assessed and made available in each country on the HarvestPlus website.

The Indian Council of Agricultural Research (ICAR) has improved the nutritional value of high yielding varieties of grains, pulses, oilseeds, vegetables, and fruits through breeding techniques [44, 45]. Special efforts were started during the 12th Plan with the development of a specific project on the Consortium Research Platform on Biofortification. 71 different varieties of rice, wheat, maize, pearl millet, finger millet, groundnut, linseed, mustard, soybean, cauliflower, potato, sweet potato, greater yam, and pomegranate have been developed as a consequence of coordinated efforts in collaboration with other national and international initiatives. Advanced

Figure 2.
Biofortified crop map. Source [39].

elite materials in considerable quantities are also in development and will be made available when the time is appropriate. The nutritional security of the country is greatly enhanced by these biofortified types. A lot of effort is put into promoting the biofortified millet cultivars.

High-quality cultivar-specific seeds that have been biofortified are developed and made available for commercial production. The Extension Division of ICAR has also introduced the Value Addition and Technology Incubation Centers in Agriculture (VATICA) and Nutri-sensitive Agricultural Resources and Innovations (NARI) special programmes to scale up the biofortified cultivars through its Krishi Vigyan Kendras (KVKs) [44].

6.7 ICAR Released some bio fortified Millets which are enlisted below:

a. Pearl millet: HHB 299 (Hybrid)

b. Compared to popular varieties' and hybrids' 45.0–50.0 ppm iron and 30.0–35.0 ppm zinc, this variety is rich in iron (73.0 ppm) and zinc (41.0 ppm). Adaptation: Kharif season in Tamil Nadu, Haryana, Rajasthan, Gujarat, and Punjab [32]. Development: CCS-Haryana Agricultural University, Hisar and ICRISAT, Patancheru as part of the ICAR-All India Coordinated Research Project on Pearl Millet. Maturity: 81 days. Grain yield: 32.7 q/ha. Dry fodder yield: 73.0 q/ha. 2017 is the release year.Pearl Millet: AHB 1200Fe (Hybrid)

c. Grain yield of 32.0 q/ha, dry fodder yield of 70.0 q/ha, maturity of 78 days, and adaptation to Kharif season in Haryana, Rajasthan, Gujarat, Punjab, Delhi, Maharashtra, and Tamil Nadu. High in iron (73.0 ppm), as opposed to popular types and hybrids that only contain 45.0–50.0 ppm. It was created by Vasantrao Naik Marathwada. For the ICAR's All India Coordinated Research Project, which will be published in 2018, Krishi Vidyapeeth in Patancheru and ICRISAT are working together to do research on pearl millet. 2018 is the project's publication year.Pearl Millet: AHB 1269Fe (Hybrid)

d. High quantities of iron (91.0 ppm) and zinc (44.0 ppm), as opposed to hybrids' (30.0–35.0 ppm) and popular varieties' (45.0–50.0 ppm) lower levels. Grain yield: 31.7 q/ha; dry fodder yield: 74.0 q/ha; maturity: 82 days; adaptation: Kharif season in Punjab, Tamil Nadu, Gujarat, Telangana, Maharashtra, and Haryana • Created by Vasantrao Naik Marathwada Krishi Vidyapeeth, Parbhani, as a component of the ICAR-All India Coordinated Research Study on Pearl Millet. The release year is 2018.Pearl Millet: ABV 04 (Open Pollinated Variety)

e. High in iron (70.0 ppm) and zinc (63.0 ppm), unlike common types and hybrids' 45.0–50.0 ppm iron and 30.0–35.0 ppm zinc • 86-day maturity • Maturity: 86 days • Grain yield: 28.6 q/ha • Dry fodder yield: 58.0 q/ha • Kharif season in Maharashtra, Karnataka, Andhra Pradesh, Telangana, and Tamil Nadu • Released in 2018 and produced by the ARS at Acharya NG Ranga Agricultural University in Ananthapuram as a component of the ICAR's All India Coordinated Research Initiative on Pearl Millet.Pearl Millet: Phule Mahashakti (Hybrid)

f. High in zinc (41.0 ppm) and iron (87.0 ppm), as opposed to hybrids' and popular types' 45.0–50.0 ppm iron and 30.0–35.0 ppm zinc, respectively. Kharif season in Maharashtra is an adaptation. Developed by the Mahatma Phule Krishi Vidyapeeth in Dhule as part of the ICAR-All India Coordinated Research Study on Pearl Millet, it has the following characteristics: • Grain yield: 29.3 q/ha • Dry fodder yield: 56.0 q/ha • Maturity: 88 days, and it was released in 2018.Pearl Millet: RHB 233 (Hybrid)

g. Instead of the hybrids'/common varieties' 45.0–50.0 ppm iron and 30.0–35.0 ppm zinc, these plants are rich in iron (83.0 ppm) and zinc (46.0 ppm). • 80-day maturity • Kharif season in Tamil Nadu, Rajasthan, Gujarat, Haryana, Madhya Pradesh, and Delhi. • Grain yield: 31.6 q/ha; dry fodder yield: 74.0 q/ha; maturity: 80 days • Released in 2019 and developed by the Sri Karan Narendra Agricultural University in Jobner as a component of the ICAR's All India Coordinated Research Initiative on Pearl Millet.Pearl Millet: RHB 234 (Hybrid)

h. High in zinc (46.0 ppm) and iron (84.0 ppm), as opposed to common types' and hybrids' 30.0–35.0 ppm and 45.0–50.0 ppm, respectively. • 81-day maturity Grain yield: 31.7 q/ha; dry fodder yield: 70.0 q/ha; maturity: 81 days; adaptation: Kharif season in Rajasthan, Gujarat, Haryana, Madhya Pradesh, Delhi, Maharashtra, and Tamil Nadu. Released in 2019 and developed by the Sri Karan Narendra Agricultural University in Jobner as a component of the ICAR's All India Coordinated Research Initiative on Pearl Millet.Pearl Millet: HHB 311 (Hybrid)

i. Grain yield of 31.7 q/ha, dry fodder yield of 72.0 q/ha, maturity of 81 days, and adaptation to Kharif season in Rajasthan, Gujarat, Haryana, Punjab, Delhi, Maharashtra, and Tamil Nadu • Released in 2020, the Pearl Millet was developed at the CCS Haryana Agricultural University in Hisar as a component of the ICAR-All India Coordinated Research Project on Pearl Millet.Finger Millet: VR 929 (Vegavathi) (Pure line variety)

j. With grain yields of 36.1 q/ha and dry fodder yields of 72.0 q/ha, this variety is rich in iron (131.8 ppm), as opposed to popular cultivars' 25.0 ppm. The cultivar, which has a 118-day maturation period, was created by Acharya NG Ranga Agricultural University in Guntur as part of the ICAR-All India programme. The Co-ordinated Small Millets Research Project is scheduled for delivery in 2020.Finger Millet: CFMV1 (Indravati) (Pure line variety)

k. This variety has 428 mg/100 g calcium, 58.0 ppm iron, and 44.0 ppm zinc, whereas standard types have 200 mg/100 g calcium, 25 ppm iron, and 16 ppm zinc. • Good for rainfed situations; Maturity: 110–115 days • Maturity: 110–115 days • Dry fodder yield: 84.4 q/ha • Grain yield: 31.1 q/ha The Kharif season is observed in the states of Andhra Pradesh, Tamil Nadu, Karnataka, Puducherry, and Odisha. The publication year is 2020.Finger Millet: CFMV 2 (Pure line variety)

l. High in calcium (454 mg/100 g), iron (39.0 ppm), and zinc (25.0 ppm), as opposed to common versions. 25 ppm iron, 16 ppm zinc, and 200 mg/100 g calcium • Good for rainfed situations, maturity: 119–121 days Grain yield was 29.5 q/ha, and dry fodder production was 86.1 q/ha. Kharif season as an adaptation in Gujarat, Maharashtra, Andhra Pradesh, and Odisha Developed by the Hill Millet

Research Station at the Navsari Agricultural University in Waghai, this research is being done as part of the ICAR-All India Coordinated Research Project on Small Millets. produced in 2020 and released.Little Millet: CLMV1 (Pure line variety)

m. High in zinc (35.0 ppm) and iron (59.0 ppm), as opposed to popular varieties' 25 ppm and 20 ppm, respectively Release year 2020; ICAR-Indian Institute of Millets Research, Hyderabad; yields of grain: 15.8 q/ha; yields of dry fodder: 55.5 q/ha; maturities: 98–102 days; suitable for rainfed conditions; adapted to Kharif season in Maharashtra, Andhra Pradesh, Telangana, Tamil Nadu, and Puducherry.

7. Components for global delivery

In order for biofortification to be extensively used and truly sustainable, several institutions must be involved in building an enabling environment. This includes adoption by the private sector, inclusion in multilaterally financed development policies and programmes, and incorporation into actual development initiatives that are being carried out on the ground, both inside and outside of target nations. This enabling environment is essential for fostering the growth of biofortified crops and supporting national actors across a range of sectors [46].

7.1 Standards and regulatory

The Food and Agriculture Organisation of the United Nations (FAO) and the World Health Organisation (WHO) jointly administer the Codex Alimentarius, the organisation that sets food standards and is acknowledged as the reference organisation by the Sanitary and Phytosanitary Agreement (SPS) of the World Trade Organisation (WTO) (FAO). There are projects in place to incorporate biofortification into these international standards and guidelines. The Codex Alimentarius is still working on defining biofortification and developing a set of guidelines for it. The widely accepted Codex reference standard, once it is adopted, aids in promoting biofortified foods and crops across borders, standardising labelling and health claims, and decreasing the prevalence of misleading claims [47].

7.2 Multi-lateral institutions

Beyond their particular investments and activities, multilateral organisations like the World Bank, the African Development Bank, the World Food Programme, and the World Health Organisation collectively have an impact on national government policymakers and operational partners. One of the World Bank's current biofortification-supporting programmes is the Multisectoral Food Security and Nutrition Project in Uganda, which is quickening the scale-up of orange sweet potatoes and iron beans. The Bank is instrumental in advancing nutrient-sensitive agricultural practises, such as biofortification, in forums like the Global Donor Forum for Rural Development. The African Development Bank's new "Banking on Nutrition" technical collaboration is implementing a multi-sectoral and integrated strategy to nutrition interventions, including the use of biofortified crops. The World Food Programme's (WFP) Buy for Progress programme, which is very interested in local purchases of biofortified crops, is forming partnerships in a number of countries. For instance, local iron bean produce is purchased and stored in WFP facilities in Rwanda in case of future disasters.

In 2017, the WHO Nutrition Guidelines Expert Advisory Group is anticipated to issue recommendations and guidelines for biofortification as a public health nutrition intervention. One phase in the process will be the publication of papers discussed in 2016 during an expert consultation held at the New York Academy of Sciences [47].

7.3 Private sectors

As a result of agricultural development programmes, more biofortified crop varieties are being published, therefore farmers need to have access to seeds from these types. Private seed companies are a natural partner in countries with robust private seed systems that reach smallholder farmers. To ensure that there would be a market for the private sector's seed and reduce the risk associated with that investment, HarvestPlus has negotiated partnerships in some cases between seed producers and interested NGOs or governments. Despite the fact that the private sector has mostly accepted hybrid crops, interest in a wider range of crops has increased as the commercial rationale for them has been established. Private sector seed businesses are brought in to assist with marketing, development, and testing of biofortified cultivars, thereby reducing time to market and establishing the groundwork for sustainability. Food processing companies are developing a significant portion of the value chain for foods manufactured from biofortified crops. Small and medium-sized firms can help increase demand even before supplies of biofortified grain and food are scaled up. For certain commodities and nations, like cassava in Nigeria, small and medium-sized food processors predominate the food value chain. The interest of multinational firms in biofortified crops is still growing, but many are already experimenting with them in their food products. These companies add to the corpus of knowledge on vitamin and mineral retention by analysing various methods of processing for minerals and vitamins [47].

7.4 NGOs

Although while private sector participation is vital for creating sustainable markets for biofortified seed and foods, NGOs nevertheless play a significant role in providing this nutrition intervention to those that are most in need. The present global relationship between World Vision and Harvest Plus serves as an example of how a leading development NGO may integrate biofortified crops into its ongoing agricultural efforts and link them to health and nutrition initiatives. Currently, Harvest Plus provides technical support while World Vision, which operates in 15 countries, leads in delivery. This kind of collaboration, where biofortified crops are incorporated into already-existing agriculture and nutrition projects or included in newly developed projects developed collaboratively, will continue to be essential to reach the most vulnerable households, which may also be the most likely to experience micronutrient deficiencies [47].

7.5 Extending beyond target nations to partnering country strategies

The government-sponsored biofortification programmes in Brazil, China, and India that are not in the target countries have received funding from Harvest Plus, their support has been extended, and they now work closely with them. Through the Harvest Plus Latin American and Caribbean (LAC) programme, which is run by the Research Corporation of the Brazilian Ministry of Agriculture, Harvest Plus provides technical support and assistance to government-led biofortification programmes in

Bolivia, Colombia, Guatemala, Haiti, Nicaragua, and Panama (EMBRAPA). Harvest Plus is also researching initiatives in a number of other nations. Such a collaborative effort is essential as biofortification gets momentum. While Harvest Plus continues to provide technical support and promote links between groups, other organisations and individuals will increasingly take the lead in delivery on the ground [39].

8. A prospective scenario: institutional leadership directing and driving mainstreaming

For biofortification to reach its full potential, it must be included as a primary activity within a number of international organisations. Three crucial elements are required, and they are as follows [47, 48]:

8.1 Supply

Agricultural research organisations, both public and corporate, now recognise high mineral and vitamin content as crucial plant breeding goals. In order to be approved for release, varieties must now meet minimal requirements for vitamins and minerals (in addition to the standard agronomic traits, such as high yield).

8.2 Policy

Many domestic and international public leaders start to recognise the significant contribution biofortification makes to maintaining and improving public health, as well as the high economic return on investment in biofortification and the legitimacy afforded by universal acclaim (especially by standards bodies).

8.3 Demand

Customers in both urban and rural areas are beginning to appreciate and demand foods with high mineral and vitamin concentrations. The secret to guaranteeing a consistent supply of biofortified crops goes beyond a breeding programme focused on biofortification, with funding committed explicitly.

9. Conclusions

Millets being an immense source of essential nutrients are dietary staple for a wide range of population both in Asian and African countries especially of those belonging to economically weaker sections of the society. Still so many developing countries are facing micro and macro nutrient deficiency. Sustainable development goals have gave more focus on nutritional aspects.

In light of newly discovered information in the genome sequences of several minor millets, now is the ideal time to utilise genomic areas determining nutritional properties in breeding programs. Because of these breeding programmes so many biofortified varierties have been introduced to combat the nutrition related issues to give the food security. Government, private sectors, NGO's have joined their hands to solve problem. As a result of recently revealed insights in the genome sequences of several minor millets, it is now time for breeding efforts to make advantage of genomic areas determining nutritional properties. Overall, millets could be promoted as a model system for the advancement of quality traits and used as a staple crop in the global economy by combining conventional and traditional breeding with the collective approach utilising all of the omics tools, including genomics, transcriptomics, proteomics, and metabolomics.

Author details

R.K. Anushree[1], Shailja Durgapal[2], Meenal[3] and Latika Yadav[4]*

1 Vasantrao Naik Marathwada Krishi Vidyapeeth, Parbhani, Maharashtra, India

2 Punjab Agriculture University, Ludhiana, Punjab, India

3 Govind Ballabh Pant University of Agriculture and Technology, Uttarakhand, India

4 Vijay Singh Pathik Government PG Degree College, Kairana, Shamli, Uttar Pradesh, India

*Address all correspondence to: drlatika27@gmail.com

References

[1] Bouis H. Reducing mineral and vitamin deficiencies through biofortification: Progress under HarvestPlus. In: Biesalski HK, Birner R, editors. Hidden Hunger: Strategies to Improve Nutrition Quality, World Review of Nutrition and Dietetics. Basel: Karger; 2018. pp. 112-122

[2] Global Nutrition Report. Nourishing the SDGs. Bristol, UK: Development Initiatives; 2017

[3] UNICEF/WHO/WB Group. Joint Child Malnutrition Estimates - Levels and Trends in Child Malnutrition. New York, Geneva, Washington, DC: UNICEF, WHO, The World Bank; 2018

[4] Saltzman A, Birol E, Bouis HE. Biofortification: Progress toward a more nourishing future. Global Food Security. 2013;**2**(1):9-17

[5] Belton PS, Taylor JRN. Sorghum and millets: Protein sources for Africa. Trends in Food Science and Technology. 2004;**15**:94-98. DOI: 10.1016/j.tifs.2003.09.002

[6] O'Kennedy MM, Grootboom A, Shewry PR. Harnessing sorghum and millet biotechnology for food and health. Journal of Cereal Science. 2006;**44**:224-235. DOI: 10.1016/j.jcs.2006.08.001

[7] Food and Agricultural Organization [FAO]. 2015. Available from: http://faostat3.fao.org/browse/Q/QC/E

[8] Yadav OP, Rai KN, Bidinger FR, Gupta SK, Rajpurohit BS, Bhatnagar SK. Pearl millet (Pennisetum glaucum) restorer lines for breeding dual-purpose hybrids adapted to arid environments. Indian Journal of Agricultural Science. 2012;**82**:922-927

[9] Yadav OP, Rai KN. Genetic improvement of pearl millet in India. Agricultural Research. 2013;**2**:275-292. DOI: 10.1007/s40003-013-0089-z

[10] Agricultural Statistics, Government of India. Pearl millet. In: Agricultural Statistics at a Glance–2014, Government of India. New Delhi: Oxford University Press; 2014. pp. 85-86

[11] Nedumaran S, Bantilan MCS, Gupta SK, Irshad A, Davis JS. Potential Welfare Benefit of Millets Improvement Research at ICRISAT: Multi Country-Economic Surplus Model Approach, Socioeconomics Discussion Paper Series Number 15. Hyderabad: ICRISAT; 2014

[12] Vijayakumari J, Mushtari Begum J, Begum S, Gokavi S. Sensory attributes of ethnic foods from finger millet (Eleusine coracana). In: Proceeding of the National Seminar on Processing and Utilization of Millet for Nutrition Security: Recent Trends in Millet Processing and Utilization. Hisar: CCSHAV; 2003. pp. 7-12

[13] Baltensperger DD. Progress with proso, pearl and other millets. In: Janick J, Whipkey A, editors. Trends in New Crops and New Uses. VA: ASHS Press; 2002. pp. 100-103

[14] Austin Kimata M, Negishi M. Geographical distribution of proso millet, Panicum miliaceum L. on iodostarch and phenol reactions; with special references to a nobalrthern propagation route into Japanese Islands. Environment Education Studies. Tokyo Gakugei University. 2002;**12**:15-22

[15] Sood S, Khulbe RK, Gupta AK, Agrawal PK, Upadhyaya HD, Bhatt JC. Barnyard millet – A potential food and

feed crop of future. Plant Breeding. 2015;**134**:135-147. DOI: 10.1111/pbr.12243

[16] de Wet JMJ, Rao KEP, Brink DE. Systematics and domestication of Panicum sumatrense (Gramineae). Journal d'Agriculture Tropicale et de Botanique Appliquée. 1983a;**30**:159-168. DOI: 10.3406/jatba.1983.3898

[17] de Wet JMJ, Rao KEP, Mengesha MH, Brink DE. Diversity in kodo millet, Paspalum scrobiculatum. Economic Botany. 1983b;**37**:159-163. DOI: 10.1007/BF02858779

[18] Hegde PS, Rajasekaran NS, Chandra TS. Effects of the antioxidant properties of millet species on oxidative stress and glycemic status in alloxan induced rats. Nutrition Research. 2005;**25**:1109-1120. DOI: 10.1016/j.nutres.2005. 09.020

[19] Saleh ASM, Zhang Q, Chen J, Shen Q. Millet grains: Nutritional quality, processing, and potential health benefits. Comprehensive Reviews in Food Science and Food Safety. 2013;**12**:281-295. DOI: 10.1111/1541-4337.12012

[20] Oberleas DP. Phytates. In: Strong FM, editor. Toxicants Occurring Naturally in Foods. Washington DC: National Academy of Sciences; 1973. pp. 363-371

[21] Gupta VP. Genetics of quality improvement. In: Gupta VP, Minnocha JL, editors. Trends in Genetical Research on Pennisetums. Ludhiana: Punjab Agricultural University; 1980. pp. 291-294

[22] Kumar A, Chauhan BM. Effect of phytic acid on protein digestibility (in vitro) and HCl-extractability of minerals in pearl millet sprouts. Cereal Chemistry. 1993;**70**:504-506

[23] Abdalla AA, El Tinay AH, Mohamed BE, Abdalla AH. Proximate composition, starch, phytate and mineral contents of 10 pearl millet genotypes. Food Chemistry. 1998;**63**:243-246. DOI: 10.1016/S0308-8146(97)00228-8

[24] Abdel Rahman SM, Babiker EE, El Tinay AH. Effect of fermentation on antinutritional factors and HCl extractability of minerals of pearl millet cultivars. Journal of Food Technology. 2005;**3**:516-522

[25] Shivaraj B, Pattabiraman TN. Natural plant enzyme inhibitors. Characterization of an unusual alpha-amylase/trypsin inhibitor from ragi (Eleusine coracana Geartn.). The Biochemical Journal. 1981;**193**:29-36. DOI: 10.1042/bj1930029

[26] Pattabiraman TN. Trypsin/chymotrypsin inhibitors from millets. Advances in Experimental Medicine and Biology. 1986;**199**:439-448. DOI: 10.1007/978-1-4757-0022-0_25

[27] Joshi BN, Sainani MN, Bastawade KB, Deshpande VV, Gupta VS, Ranjekar PK. Pearl millet cysteine protease inhibitor. Evidence for the presence of two distinct sites responsible for anti-fungal and anti-feedant activities. European Journal of Biochemistry. 1999;**265**:556-563. DOI: 10.1046/j.1432-1327.1999.00764

[28] Muthamilarasan M, Prasad M. Advances in Setaria genomics for genetic improvement of cereals and bioenergy grasses. Theoretical and Applied Genetics. 2015;**128**:1-14. DOI: 10.1007/s00122-014-2399-3

[29] O'Kennedy MM, Crampton BG, Ozias-Akins P. Pearl millet. In: Kole C, Hall TC, editors. Compendium of Transgenic Crop Plants: Transgenic Cereals and Forage Grasses. Vol. 1.

Chichester: John Wiley & Sons; 2009. pp. 177-190

[30] Tsehaye Y, Berg T, Tsegaye B, Tanto T. Farmers' management of finger millet (Eleusine coracana L.) diversity in Tigray, Ethiopia and implications for on-farm conservation. Biodiversity and Conservation. 2006;**15**:4289-4308. DOI: 10.1007/s10531-005-3581-3

[31] Reddy INBL, Reddy DS, Narasu ML, Sivaramakrishnan S. Characterization of disease resistance gene homologues isolated from finger millet (Eleusine coracana L. Gaertn). Molecular Breeding. 2011;**27**:315-328. DOI: 10.1007/s11032-010-9433

[32] Yadava DK, Choudhury PR, Hossain F, Kumar D, Sharma TR, Mohapatra T. Biofortified Varieties: Sustainable Way to Alleviate Malnutrition. New Delhi: Indian Council of Agricultural Research; 2020. p. 19

[33] Global Food Policy Report. Washington, DC: International Food Policy Research Institute; 2016

[34] Menon P, Headey D, Avula R, Nguyen PH. Understanding the geographical burden of stunting in India: A regression-decomposition analysis of district-level data from 2015-16. Maternal and Child Nutrition. 2017;**14**:e12620. Available from: https://doi.org/10.1111/mcn.12620

[35] National Family Health Survey-4. Ministry of Health and Family Welfare. Government of India; 2015

[36] Harvestplus.org. 2020. Available from: https://www.harvestplus.org/where-we-work/india

[37] Bouis HE, Hotz C, McClafferty B, Meenakshi JV, Pfeiffer WH. Biofortification: A new tool to reduce micronutrient malnutrition. Food and Nutrition Bulletin. 2011;**32**:S31-S40. DOI: 10.1177/15648265110321S105

[38] Hotz C, McClafferty B. From harvest to health: Challenges for developing biofortified staple foods and determining their impact on micronutrient status. Food and Nutrition Bullettin. 2007;**28**(2):S-271-S-279

[39] HarvestPlus. Disseminating Orange-Fleshed Sweet Potato: Findings from a HarvestPlus Project in Mozambique and Uganda. Washington, DC: HarvestPlus; 2010

[40] Trijatmiko KR, Duenas C, Tsarkirpaloglou N, et al. Biofortified indica rice attains iron and zinc nutrition dietary targets in the field. Scientific Reports. 2016;**6**:19792

[41] Wesseler J, Zilberman D. The economic power of the Golden Rice opposition. Environment and Development Economics. 2014;**19**:724-742

[42] Paltridge NG, Milham PJ, Ortiz-Monasterio JI, Velu G, et al. Energydispersive X-ray fluorescence spectrometry as a tool for zinc, iron and selenium analysis in whole grain wheat. Plant and Soil. 2012a;**361**:261-269

[43] Paltridge NG, Palmer LJ, Milham PJ, Guild GE, et al. Energy-dispersive X-ray fluorescence analysis of zinc and iron concentration in rice and pearl millet grain. Plant and Soil. 2012b;**361**:251-260

[44] Yadava DK, Hossain F, Mohapatra T. Nutritional security through crop biofortification in India: Status & future prospects. Indian Journal of Medical Research. 2018;**148**:621-631. DOI: 10.4103/ijmr.IJMR_1893_18

[45] Yadava DK, Choudhury PR, Hossain F, Kumar D, Mohapatra T. Biofortified Varieties: Sustainable Way to Alleviate Malnutrition (Second Edition). New Delhi: Indian Council of Agricultural Research; 2019. 44p

[46] Bouis HE, Saltzman A. Improving nutrition through biofortification: A review of evidence from HarvestPlus, 2003 through 2016. Global Food Security. 2017;**12**:49-58

[47] Babu R, Palacios-Rojas N, Gao S, Yan J, Pixley K. Validation of the effects of molecular marker polymorphisms in LcyE and CrtRB1 on provitamin a concentrations for 26 tropical maize populations. Theoretical and Applied Genetics. 2013;**126**(2):389-399

[48] Swamy BPM, Rahman MA, Inabangan-Asilo MA. Advances in breeding for high grain zinc in rice. Rice. 2016;**9**:49. DOI: 10.1186/s12284- 016-0122-5

Chapter 5

Agronomic Biofortification of Millets: New Way to Alleviate Malnutrition

Sreenivasareddy Kadapa, Alekhya Gunturi, Rajareddy Gundreddy, Srikanth Reddy Kalwala and Uday Bhaskar Mogallapu

Abstract

Biofortification or biological fortification refers to nutritionally enhanced food crops with increased bioavailability to the human population that are developed and grown using modern bio-technology techniques, conventional plant breeding, and agronomic practices. Our agricultural system has not been designed to promote human health; instead, it only focuses on increasing grain yield and crop productivity. This approach has resulted in a rapid rise in micronutrient deficiency in food grains, thereby increasing micronutrient malnutrition among consumers. Biofortification is a one-time investment and offers a cost-effective, long-term, and sustainable approach in fighting hidden hunger because once the biofortified crops are developed, there are no costs of buying the fortificants and adding them to the food supply during processing. Agronomic biofortification methods requires physical application of nutrients to temporarily improve the nutritional and health status of crops and consumption of such crops improves the human nutritional status. Soil and plant are managed by agronomic interventions. For the biofortification initiative to be successful, farmers use micronutrient fertilizers to fortified cultivars must get marketing support. Besides challenges the biofortification of millets have a promising future in combating the problem of malnutrition.

Keywords: biofortification, malnutrition, agronomic biofortification, fertilization approaches, millets biofortification

1. Introduction

According to the United Nations Food and Agriculture Organization, 780 million of the world's estimated 792.5 million malnourished people reside in developing nations [1]. Additionally, despite increasing food crop production, around 2 billion people worldwide experience "hidden hunger," which is brought on by a lack of vital micronutrients in the daily diet [2, 3]. In addition, there is growing concern over nutrition. Until now, the primary goals of our agricultural system have been to boost crop productivity and grain yield, not human health. This strategy has caused a sharp increase in the lack of some micronutrients in dietary grains, which has increased micronutrient malnutrition among consumers. Agriculture is currently transitioning from producing more food crops in greater quantities to generating enough nutrient-rich crops. This will aid in the battle against "hidden hunger" or "micronutrient malnutrition," particularly in underdeveloped and poorer nations whose diets are predominately composed of micronutrient-poor staple foods [4]. Since the green revolution, there has been a huge rise in the production of food crops. But the nutrient content of crops could not keep up with the growing demand from the population. Malnutrition problems have become worse as a result of a lack of a balanced diet, especially in developing countries.

Following the adoption of the Millennium Development Goals (MDGs) and then the Sustainable Development Goals, malnutrition—the monster of hidden hunger—has already attained the status of being of the utmost significance (SDGs). Every region of the world is affected by the issue of malnutrition. There are 2 billion or so malnourished individuals in the world, according to reports [5]. In this world, over 850 million of people are affected negatively by undernourishment [6]. In low-income nations like Africa, where Ca (54% of the continent's population), Zn (40%), Se (28%), I (19%), and Fe (5%), there is a considerable risk for micronutrient deficiencies [7]. In poor nations, malnutrition mostly affects mothers and small children in many ways. A key strategy for lowering the prevalence of malnutrition worldwide is biofortification of different crop varieties offers a sustainable and long-term solution in providing micronutrients-rich crops to people. The terms "biofortification" and "biological fortification" refers to nutrient-enhanced food crops that are produced and grown utilizing contemporary bio-technology approaches, traditional plant breeding, and agronomic practices. Furthermore, biofortified crops with increased bioavailable concentrations of essential micronutrients are deployed to consumers through traditional practices used by agriculture and food trade which therefore provides a feasible way of reaching undernourished and low income group families with limited access to diverse diets, supplements, and fortified foods. These crops also have greater bioavailability to the human population [8]. Biofortification is an upcoming, promising, cost-effective, and sustainable technique of delivering micronutrients to a population that has limited access to diverse diets and other micronutrient interventions. Agronomic biofortification, the practice of increasing the micronutrient content of food crops through agronomic techniques. We can quickly, safely, and economically increase the amount of iron, zinc, and other micronutrients in our diet. Contrary to molecular/genetic methods, agronomic biofortification is done on current crop type to improve the product's consumer acceptability. Major food crops, unfortunately, are poor suppliers of the micronutrients necessary for healthy human growth. The biofortified food crops, particularly the cereals, legumes, vegetables, and fruits, are giving the targeted people enough micronutrients. Although transgenic research is given more attention, breeding has a significantly higher success rate and acceptance rate. In spite of the difficulties, biofortified crops have a promising future in the fight against hunger. Poor people's purchasing ability, access to markets and healthcare systems, and ignorance about the long-term health advantages of these vitamin supplements are further barriers [9, 10]. The development of biofortified crops eliminates the need to purchase fortificants and add them to the food supply during processing, making biofortification from an economic perspective a one-time investment that provides a cost-effective, long-term, and sustainable method of addressing hidden hunger [11–14]. A substantial population rise in the developing world is also possible in the next decades, and combined with changing climate circumstances, ensuring food security will be more difficult [15, 16].

Since the majority of people consume a plant-based diet, nutritional security is essential to enhancing the health of the global population. The main source of the nutrients needed for healthy growth and development is plants. But due to their reliance on grain products, half of the world's population, mainly those from Asia and Africa, suffer from nutritional deficiencies [14, 17, 18].

2. Significance of millets and biofortification

Millets, which have an average annual production of 14.2 and 12.4 million tonnes, are the second-largest source of calories after cereal grains in resource-limited nations in Asia and Africa [19, 20]. India is the world's top producer of millets, as depicted in **Figures 1** and **2** making up around 80% of the total production [22] as mentioned in **Table 1**. Millets, which include pearl millet, finger millet, foxtail millet, proso millet, barn yard millet, kodo millet, and little millet, are frequently referred to as "small seeded grasses".

Pearl millet accounts for major share of the output of the millets [30–33]. Millets are therefore consumed as multi-grains to reap the collective health benefits of nutrients. Due to their high levels of proteins, dietary fibers, iron, zinc, calcium, phosphorus, potassium, vitamin B, and vital amino acids, millets are nutritionally superior to wheat and rice [34, 35]. However, anti-nutrients such as phytates, polyphenols, and tannins decrease the bioavailability of minerals by chelating multivalent cations including Fe^{2+}, Zn^{2+}, Ca^{2+}, Mg^{2+}, and K^{+} [36–40]. Additionally, the digestibility of millet grains is impacted by high levels of protease and amylase inhibitors [41–43] as shown below in the **Table 2**. Millets now have an orphan status in terms of worldwide economic significance due to the prevalence of antinutritional

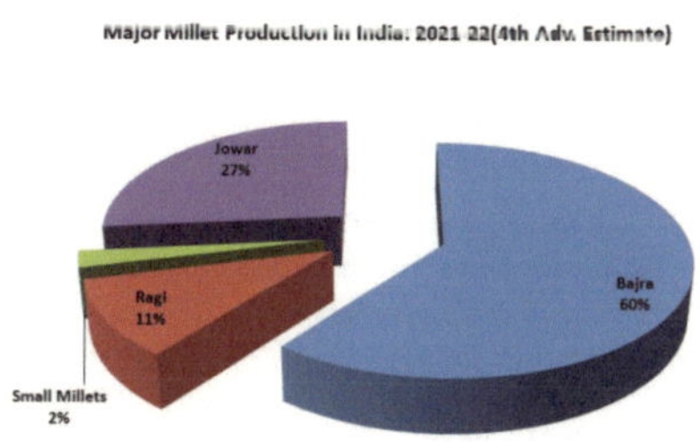

Figure 1.
Major millets production in India.

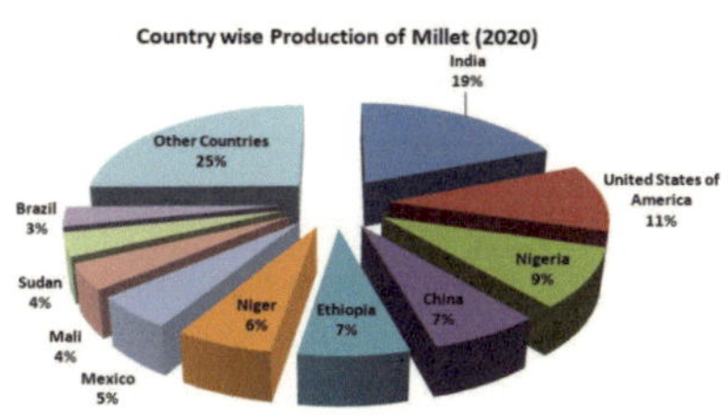

Figure 2.
Production share of millets country wise [21].

Millet	Cultivation	Reference
Foxtail millet [*Setaria italica (L.) Beauv*]	Cultivated for food in semi-arid tropics of Asia and as forage in Europe, North America, Australia, and North Africa	[23]
Finger millet [*Eleusine coracana (L.) Gaertn*]	As the primary food for rural populations of East and Central Africa and southern India	[24]
Proso millet (*Panicum miliaceum L.*)	Cultivated in drier regions of Asia, Africa, Europe, Australia, and North America	[25, 26]
Barnyard millet (*Echinochloa spp.*)	Cultivated in India, China, Japan, and Korea for food as well as fodder	[27]
Kodo millet (*Paspalum scrobiculatum*)	Native to the tropical and sub-tropical regions of South America and domesticated in India 3000 years ago	[28]
Little millet (*Panicum sumatrense*)	Domesticated in the Eastern Ghats of India occupying a major portion of diet among the tribal people and spread to Sri Lanka, Nepal, and Myanmar	[29]

Table 1.
Millets and their cultivation in this world.

Millet	Nutrition Content	Reference
Pearl millet	Rich in Fe, Zn, and lysine (17–65 mg/g of protein) compared to other millets. Total phenolic contents reported are 168 mg/100 g (pearl millet) and ferulic acid equivalents in the soluble phenolic fraction. Total flavonoid contents 49 mg/100 g (pearl millet) catechin equivalents in the soluble phenolic fraction.	[44, 45] [46]
Foxtail millet	High amount of protein (11%) and fat (4%). The protein fractions are represented by albumins and globulins (13%), prolamins (39.4%), and glutelin's (9.9%). It is thus recommended as an ideal food for diabetics. It also contains significant amounts of potential anti-oxidants like phenols, phenolic acids, and carotenoids	[35, 47]
Finger millet	Rich in Fe, Zn, and lysine (17–65 mg/g of protein) compared to other millets. Total phenolic contents reported are 168 mg/100 g (pearl millet) and ferulic acid equivalents in the soluble phenolic fraction. Total flavonoid contents have been reported as 203–228 mg/100 g (finger millet), catechin equivalents in the soluble phenolic fraction.	[35, 48, 49] [49, 50] [46]
Proso millet	Rich in Fe, Zn, and lysine (17–65 mg/g of protein) compared to other millets Total phenolic contents reported are 168 mg/100 g (pearl millet) and ferulic acid equivalents in the soluble phenolic fraction. Total flavonoid contents have been reported as 140 mg/100 g (proso millet) catechin equivalents in the soluble phenolic fraction.	[35] [46]
Barn yard millet	Functional constituents' viz. g-amino butyric acid (GABA) and b-glucan, used as anti-oxidants and in reducing blood lipid levels.	[51, 52]
Kodo millet	High magnesium content (1.1 g/kg dry matter).	
Overall view of millets	They contain health promoting phenolic acids and flavonoids, that play a vital role in combating free-radical mediated oxidative stress and in lowering blood glucose levels	[34, 46, 50] [53–56]

Table 2.
Millets as a rich source of nutritional contents.

forces. When compared to other cereal crops, millets have significantly more genetic diversity for important mineral elements including iron, zinc, and calcium [57]. Additionally, millets are pest- and disease-resistant plants that can withstand drought [58] and provide effective crop insurance in underdeveloped nations [59, 60]. A food-based strategy called biofortification puts nutrient-dense crops at the doorsteps of underprivileged communities in order to combat nutritional hunger [61]. The Harvest Plus-Consultative Group for International Agricultural Research (CGIAR) Micronutrients project's Biofortification Challenge Program (BCP) has primarily targeted three crucial micronutrients (Fe, Zn, and vitamin A) in seven major staple crops, namely rice, beans, cassava, maize, sweet potato, pearl millet, and wheat [62]. In this situation, millets biofortification may offer an effective means of ensuring the nutritional security of the world's 8 billion people. By outlining the prospects and difficulties for enhancing the bioavailability of macro and micronutrients, we may explore the methods for accelerating biofortification in millets.

3. Necessity for biofortification

Our bodies require minute amounts of vitamins, and minerals and micronutrients. However, they have a significant influence and their deficiencies lead to major health problems such as chronic illnesses and stunting, weakened immunological and reproductive systems, and a decline in our physical and mental capabilities (WHO). Each year, more than 20 million people die from micronutrient deficiency, affecting more than 2 billion people [63, 64]. It is also known as "hidden hunger." Ten main causes of sickness and disease in low-income nations, [65] of which Zn and Fe deficits rank fifth and sixth and are largely persistent. The most vulnerable groups to micronutrient deficiencies are children and women. According to WHO estimates, malnutrition, particularly a lack of micronutrients, killed approximately 6.3 million children under the age of 15 in 2017 and 5.4 million of them were under the age of five [66]. This is mostly caused by inadequate protein consumption, a lack of access to high-quality meals rich in micronutrients like iodine, iron, and zinc, or a repetitive eating pattern. Children who were stunted in the mother's womb due to the expectant mother's poor consumption of micronutrient-enriched foods. A major worldwide issue for humanity, malnutrition is believed to impact more than half of the world's population. Traditionally, pharmaceutical supplementation and industrial fortification have been key strategies for addressing nutritional concerns. But these things are low reachability to poor income countries sometimes they reluctant to intakes of this tablet. So, the efficiencies of these strategies are low. Therefore, Biofortification was presented as a novel step. It is the act of breeding nutrients into food crops and is a reasonably cheap, long-term method of enhancing micronutrient delivery. This tactic not only lowers the number of people who are extremely malnourished and require supplemental therapy, but it also helps those people retain their improved nutritional status. Additionally, biofortification is a workable solution for rural people who are in poverty and might not have access to commercially market fortified meals and supplements. They prefer cereal-based foods, which are lower in protein and vitamins, and the soils in this area are depleted in zinc (50%) and iron (30%) and iodine, with the majority of the soil being damaged by alkalinity and salt problems [67].

Millets are advised for the health of newborns, nursing mothers, the elderly, and recovering patients. The grains are regarded as "gluten-free" because they gradually release sugar into the bloodstream [68]. Millets are favored as dietary items for persons with diabetes and cardiovascular disorders because of their high fiber and protein content [69].

For biofortification to be successful, the following three issues must be resolved: A biofortified crop must meet the following criteria:

i. It must provide a high yield and be profitable for the farmer;

ii. It must be efficient and successful in decreasing micronutrient deficiencies in people;

iii. It must be accepted by both farmers and consumers in the target regions [70].

4. Techniques for biofortification

The key processes or procedures for biofortified crops are listed below.

1. Agricultural practices: Increasing the quantity of micronutrients in plants produced in soils lacking in those nutrients requires the use of fertilizers.

2. Traditional plant breeding: This entails utilizing conventional breeding techniques to generate enough genetic variety in agricultural plants, for instance, to obtain high vitamin content. To develop nutrient-rich plants and other desirable features, it entails mating several varieties over numerous generations. In India, this method is the only one used to generate bio-enriched plants.

3. Genetic engineering is introducing DNA into an organism's genome to produce new or changed properties, such as traits and introducing disease resistance.

5. Methods employed for biofortification

Biofortification has been promoted as a long-term alternative to standard treatments since they are ineffectual for boosting mineral nutrition. The process of biofortification raises the mineral content and bioavailability of staple crop edible sections. While the latter can be achieved by agronomic intervention, plant breeding, or genetic engineering, mineral bioavailability can only be affected by these two methods as shown below in **Tables 3** and **4** which describes about varieties, nutritional aspects and their adaptation.

Despite millets' higher quality, India has only given pearl millet the top priority when it comes to crops for iron biofortification. Therefore, there is a lot of room to use the minor millets for biofortification. There are two ways to accomplish biofortification in millets:

1. By boosting the accumulation of nutrients in milled grains, and.

2. By lowering the antinutrients to enhance the bioavailability of minerals as discussed in **Table 5** with development of biofortified varieties.

Sorghum	Iron	Released	India: ICSR 14001, ICSH 14002 Hybrids: ICSA 661 × ICSR 196, ICSA 318 × ICSR 94 ICSA 336 × IS 3760	ICRISAT, Harvest Plus
	Iron	Released	Nigeria: 12KNICSV(Deko)-188 112KNICSV-22 (Zabuwa)	ICRISAT, Harvest Plus
	Iron, zinc, beta-carotene	Research		[71]
Pearl Millet	Iron and zinc	Released	India: Dhanashakti Hybrid, ICMH 1201 (Shakti-1201)	ICRISAT, Harvest Plus
	Iron and zinc	Research		[72, 73]

Table 3.
Millets breeding for improving lives of million people around the world.

Varieties	Nutritive value	Adaptation zone/state	Season of cultivation	Grain yield
Dhanshakti/ ICTP 8203 Fe	Fe: 71 ppm, Zn: 40 ppm	Maharashtra, Karnataka, Telangana, Uttar Pradesh, Haryana & Rajasthan.	Kharif	2.21 t/ha
Shakti-1201/ ICMH 1201	Fe: 75 ppm, Zn: 40 ppm	Maharashtra & Rajasthan.	Kharif	3.6 t/ha
HHB-299	Fe: 73 ppm Zn: 41 ppm	Haryana, Rajasthan, Gujarat, Punjab, Delhi, Maharashtra & Tamil Nadu	Kharif	3.27 t/ha
AHB-1200Fe (Hybrid)	Fe: 73 ppm	Haryana, Rajasthan, Gujarat, Punjab, Delhi, Maharashtra &Tamil Nadu.	Kharif	3.2 t/ha

Table 4.
Millet varieties, nutritional aspects and their adaptation in various states of India.

Crop	Varieties	References
Pearl millet	HHB-299 (73.0 ppm Fe & 41.0 ppm Zn), AHB 1200 Fe, (77 ppm Fe & 39 ppm Zn), AHB 1269 Fe (91.0 ppm Fe 43.0 ppm Zn), ABV-04 (70.0 ppm Fe & 63.0 ppm Zn), Phule Mahashakti (87 ppm Fe & 41 ppm Zn), RHB-233(83 ppm Fe & 46 ppm Zn), RHB-234 (84 ppm Fe & 46 ppm Zn), HHB-311 (83 ppm Fe & 39 ppm Zn)	[75, 76]
Sorghum	Zn concentration in grain for mean of parent (IS2248 × IS 20843) was 55.46 ppm. Fe concentration in grain for mean of parent (ICSB 52 × SPV 1359) was 50.17 ppm	[77]
Finger millet	VR-929 (Vegavathi) (131.8 ppm Fe), CFMV-1 (Indravati) (58 ppm Fe & 44 ppm Zn), CFMV-2 (25 ppm Zn & 39 ppm Fe)	[76]
Little millet	CLMV-1 (59 ppm Fe & 35 ppm Zn)	

Table 5.
Genetic biofortification through identification/development of biofortified varieties of different crops [74].

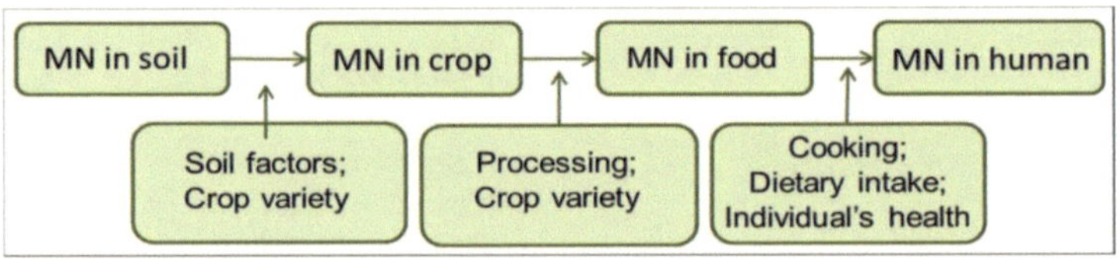

Figure 3.
Schematic overview of micronutrient (MN) pathway from soil to humans and that influence MN bioavailability to the next level [78].

This study emphasizes the value of millet germplasm characterization for creating biofortified cultivars and the application of omics techniques to increase grain-nutrient density. We highlight the use of genetic engineering and genome editing technologies to promote nutrient accumulation in edible sections and to prevent the production of anti-nutrients, following the example of other cereal crops as shown above **Figure 3**.

6. Agronomic approaches for biofortification

The most significant contribution to human health and prevention is probably an adequate and balanced diet that provides the energy routes, vital amino acids (lysine, methionine), vitamins (A, B, C, D and E), minerals, folic acids, and ionic elements (Fe, Zn, I, and Se). Targeted administration of soluble inorganic fertilizers to the roots or to the leaves is used when crops are produced in soils where mineral elements become instantly unavailable in the soil and/or not rapidly translocated to edible tissues. Agronomic biofortification is easy and cheap, but it requires specific consideration when it comes to nutrient supply, application technique, and environmental impact as discussed in **Table 6**. Hence, in certain situations, are less cost-effective. The success of Se fertilization of crops in Finland [84], zinc fertilization in Turkey [85], and I fertilization in irrigation water in China serve as examples of how mineral fertilizers might be used in developed countries [86]. In addition to fertilizers, nutrient mobility from the soil to the edible sections of plants can be improved by using

Crop	Treatment	Improvement in Zn and Fe concentration	Reference
Sorghum	Combined application of 30 kg S ha^{-1} through gypsum, 0.5 kg B ha^{-1} through borax and 10 kg Zn ha^{-1} through $ZnSO_4$	Increase in Zn concentration in grain by 7 mg kg^{-1} grain	[79]
Finger millet	Foliar sprays of 0.2% $ZnSO_4$ and Zn-EDTA twice at 30 and 60 days after sowing	Increase in total Zn and Fe uptake by 149 g ha^{-1} and 1497 g ha^{-1} in case of $ZnSO_4$ spray and 279 g ha^{-1} and 1862 g ha^{-1} with Zn-EDTA spray,	[80]
Sorghum	Mycorrhiza + Bacteria Research FYM + biofertilizer Research		[81–83]

Table 6.
Contribution of agronomic bio-fortification in increasing the grain Zn and Fe concentrations and uptake in different crops [74].

soil microbes that promote plant development. To boost the Phyto availability of mineral elements, soil microbes from the genera Bacillus, Pseudomonas, Rhizobium, and others can be used [87, 88]. When nitrogen is scarce, the N_2-fixing bacteria are crucial for enhancing crop output [89]. Numerous crops have mycorrhizal fungus attached to them, which can emit organic acids, siderophores, and enzymes that can break down organic molecules and raise the mineral content of edible product [87, 90].

A good technique for supplementing micronutrient powders and promoting dietary diversity is agronomic biofortification, which is the process of increasing the density of nutrients, vitamins, and minerals in a crop by using suitable agronomic practices.

The following are the main benefits of agronomic biofortification:

i. It is applied to crop cultivars that farmers are currently using and whose output is well-accepted by consumers.

ii. Enhanced micronutrient concentration in grain and other crop portions may be obtained in the same year.

iii. When the foliar treatment is used, very little micronutrient is required.

iv. New seed does not require investment.

v. For poor nations, agronomic biofortification usually results in a win-win situation.

6.1 Agronomic biofortification in sorghum

Sorghum crop often suffers from the challenge of growing in nutrient poor and contaminated soil. The nutrient profile has been promoted by the application of fertilizers (both organic and inorganic) that have an additive effect on the yield. Researchers have intended to improve the nutrient uptake and alter the metabolic profile of sorghum by using the combination of plant growth-promoting bacteria and arbuscular mycorrhizal fungi (AMF) [81, 82]. Also, the inoculation of *Azospirillum* alone and in combination with phosphate-solubilizing bacteria increased sorghum grain yield and protein content by improving the status of phosphorous and nitrogen in the soil [83].

6.2 Agronomy biofortification through fertilization techniques

We are unaware of other studies that similarly quantified the direct impact of agronomic biofortification on dietary intake of micronutrients on human health. Even though it is shown that agronomic biofortification has the potential to increase micronutrient contents in crops, literature connecting these enhanced concentrations to micronutrient bioavailability, dietary intake and human health are scarce [91]. Such studies do exist on genetically biofortified crops, such as in the case of Indian schoolchildren consuming iron biofortified pearl millet [92]. Modeled estimations have been made on the potential of agronomic biofortification using agronomic and dietary data [93] proposed that future study on micronutrient bioavailability, including metabolic pathways that impact absorption and the health benefits of various chemical forms of micronutrients, is necessary to further establish the legitimacy of agronomic biofortification. The rate of adoption at the stakeholder level is quicker because fertilization is associated with the economy in both the short- and long-term [94], which is evident in both the rate of micronutrient application and the usage of micronutrient-fortified fertilizers [95]. To define the potential and essential circumstances for agronomic biofortification to improve human health, systematic study is needed.

6.3 Agronomic biofortification compared with other interventions

When compared to other intervention strategies like genetic biofortification, food fortification, supplementation, and dietary diversification, the question of whether agronomic biofortification is an efficient, workable, and sustainable approach to addressing micronutrient deficiencies still needs to be answered. Rarely do economic evaluations take agronomic biofortification into account when comparing the relative effects of various initiatives on nutrition. In comparison to other treatments, genetic biofortification is more economical over time than food fortification, supplementation, or dietary diversification since it only needs one breeding investment period [96, 97]. In addition to genetic biofortification (breeding), which is viewed as a more permanent technique, agronomic biofortification is frequently perceived as a temporary option to boost micronutrient availability [98, 99]. Breeding, according to [100] is the only agricultural intervention that may increase the nutritional value of staple crops in low-income countries since farmers with limited resources do not have access to or can pay fertilizers. Dietary diversity, according to the CGIAR biofortification programme, is the best sustainable option, yet those who are most at risk frequently cannot afford various foods. According to [101] concentrated metropolitan regions are most suited for supplements and food diversification programmes, whereas rural people are best served through agronomic biofortification.

7. Developments in biofortification

As of 2018, 6.7 million farm families throughout the world produced biofortified crops, and these goods undoubtedly end up in meals. More than 300 different cultivars of crops, including rice, wheat, maize, cassava, orange sweet potato, potato, lentil, beans, cowpea, banana, and plantain, have been made available so far in 30 different countries [102]. To populate biofortified crops and create an enabling environment, a number of institutions including the.

1. Food Policy Research Institute (IFPRI),

2. Biotechnology Industry Research Assistance Council (BIRAC),

3. Bill and Melinda Gates Foundation (BMG Foundation), and

4. Indian Agricultural Research Institute (IARI) must collaborate.

An environment like this includes, among other things, designing new development policies and agendas that take into account the programmes currently being implemented on the ground, recognizing biofortification among global regulatory agencies, collaborating between agencies from various sectors, encouraging private players to play an active role, and more. In order to promote a single, integrated dialog on standards and governance and to provide society the maximum return on investment feasible, CGIAR will continue to use its diverse network of international organizations, research institutes, and civil society groups throughout the world. One of them, Harvest Plus, is in charge of the biofortification initiative, which it will enable over the next years with the primary participation of local governments [102].

There is considerable evidence that eating these biofortified cereals with added macro- and micronutrients helps reduce malnutrition in underdeveloped nations. Micronutrient deficits in India have become worse due to the growth of high yielding cultivars, numerous cropping methods requires attention [103] and rising soil degradation [104]. Agronomic (raising micronutrients by soil amendments or foliar spray), biofortification is the intentional application of mineral fertilizers (such as enriched manures) to crops in order to raise the concentration of a target mineral in edible crop components and hence improve dietary intake of the target mineral [94] conventional breeding (which includes induced mutagenesis), and recombinant DNA technology (genetic engineering, GM) are the three biofortification procedures that have been found [105]. Both the impacts on the nutritional value of these small grains and the effects on the end-use functional qualities will be studied.

8. Advantages of biofortification

Similar to the Green Revolution, a hunger-eradication initiative was launched in India. Due to the Green Revolution, the nation is producing more edible grains and is now nearly self-sufficient. To make sure that the populace is ingesting adequate calories, the government is putting many programmes and procedures into place. The improvement of the diet's nutritional content, however, is the present priority. Even when they consume "adequate food," many people may not acquire enough nutrients. The issue of "hidden hunger" is the outcome. With Prime Minister Narendra Modi's recent support of locally grown foods as a long-term, economical solution to malnutrition, the Government of India (GoI) is encouraging biofortification. The availability of several biofortified crops in India, such as iron pearl millet, zinc wheat, zinc rice, zinc sorghum, and iron/zinc lentils, helps address the country's micronutrient deficits by raising dietary levels of iron and zinc.

By utilizing distinct strategies, biofortification gives developing nations several benefits. In order to boost crops, research and initiatives like Harvest Plus are concentrating on the micronutrients iron (Fe), zinc (Zn), and vitamin A, which are considered by the World Health Organization to be the most scarce micronutrients. These common crops can be found everywhere and do not require specific management because it is possible to enhance the yield without compromising the crop's productivity. It can even lead to better development and larger yields because the majority of the target minerals are crucial for the plant's own nutritional requirements and may help the plant tolerate environmental stress. It is practiced on crop cultivars that farmers are already growing and have good production acceptability, to increase the micronutrient content of grain and other agricultural components in the same year. When the foliar treatment is used, very little micronutrient is required. Further, no investment is required. The agronomic techniques that we can use to boost the concentration of nutrients in edible parts. Helps in maintaining the physical, chemical, and biological characteristics of the soil with Integrated Nutrient management practices.

9. Potential ways for agronomic biofortification

a. Testing existing varieties for receptivity to micronutrients,

b. Creating biofortified cultivars to address micronutrient deficits,

c. Micronutrient need and fertilization in difficult soils,

d. Importance of Zn and Fe fertilization in disease resistance,

e. Micronutrient deficiencies and their management in rainfed agriculture,

f. Soil micronutrient status across cropping systems and location and utilization in policy orientation,

g. Zinc and Fe fertilization approaches for growth and yield enhancement

h. Methods for boosting the bioavailability of micronutrients in key food items,

i. Research on improving the usage efficiency of applied mineral micronutrient fertilizers

j. The utilization of microbial inoculations and rhizosphere manipulation through management methods and input addition are potential ways to open the doors of this pool, which has not yet been used.

k. Along with micronutrients, crops that have been agronomically biofortified to address protein, vitamin A, and folic acid deficits also need to be given similar priority [74].

10. Constraints in agronomic biofortification

The following difficulties arise while improving crop characteristics through agronomic biofortification: Micronutrients like iron, zinc, copper, etc. have relatively poor utilization efficiencies (1–5%), which restricts the absorption of applied micronutrients by plants.

a. Farmers' timely access to micronutrient fertilizers

b. Genetic restrictions

c. Difficulty in raising public awareness

d. Lack of knowledge

e. Post-harvest processing losses

The simplest approach of biofortification is the use of micronutrient-enriched fertilizers. But because of variations in mineral mobility, mineral accumulation across plant species, and soil compositions in the specific geographic location of each crop, the efficacy of agronomical biofortification is very varied. Agronomic approaches provide a short-term solution compared to breeding approaches.

11. Conclusion

The success of agronomic biofortification depends on the bioavailability of micronutrients along the entire pathway from soil to plant, food, and the human body. Since there are few studies linking the use of micronutrient fertilizer to improved human health, the effectiveness and utility of agronomic biofortification to treat human micronutrient deficiencies. We recommend the creation of research and pilot-scale fertilization programmes to bridge the knowledge gap on the relationship between the application of micronutrient-enriched fertilizer to crops and dietary micronutrient intake and absorption in consumer's bodies. In the short term, agronomic approaches are the most important sustainable techniques of biofortification. Besides these challenges, biofortified crops hold a very bright future as these have the potential to remove micronutrient malnutrition among billions of poor people, especially in developing countries. It is well established that biofortification is a promising, cost effective, agricultural strategy for improving the nutritional status of malnourished populations throughout the world. The generation of biofortified food crops with improved nutrient contents such as increases in iron, zinc, Se, and provitamin A content are providing sufficient levels of these and other such micronutrients that are frequently lacking in the diets of the developing and developed world. To achieve this, collaboration between plant breeders, nutrition scientists, genetic engineers, and molecular biologists is essential. Besides these challenges, biofortified crops hold a very bright future as these have the potential to remove micronutrient malnutrition among billions of poor people, especially in the developing countries. The concept of biofortification should be viewed as the soil–plant–animal–human as a continuum rather than working on any one component in the food chain.

Author details

Sreenivasareddy Kadapa[1], Alekhya Gunturi[1], Rajareddy Gundreddy[2], Srikanth Reddy Kalwala[1*] and Uday Bhaskar Mogallapu[3]

1 Division of Agronomy, ICAR-Indian Agricultural Research Institute, New Delhi, India

2 Division of Entomology, ICAR-Indian Agricultural Research Institute, New Delhi, India

3 Department of Agricultural Extension, S.V. Agricultural College, Tirupati, India

*Address all correspondence to: ksreddyagro@gmail.com

References

[1] McGuire S, FAO, IFAD, WFP. The state of food insecurity in the world 2015: Meeting the 2015 international hunger targets: Taking stock of uneven progress. Rome: FAO. Advances in Nutrition. 2015;**6**(5):623-624

[2] Hodge J. Hidden hunger: Approaches to tackling micronutrient deficiencies. In: Gillespie S, Hodge J, Yosef S, Pandya-Lorch R, editors. Nourishing Millions: Stories of Change in Nutrition. Washington: International Food Policy Research Institute (IFPRI); 2016. pp. 35-43

[3] Muthayya A, Rah JH, Sugimoto JD, Roos FF, Kraemer K, Black RE. The global hidden hunger indices and maps: An advocacy tool for action. PLoS One. 2013;**8**(6):67860

[4] Khush GS, Lee S, Cho JI, Jeon JS. Biofortification of crops for reducing malnutrition. Plant Biotechnology Reports. 2012;**6**:195-202

[5] Tulchinsky T. Micronutrient deficiency conditions: Global health issues. Public Health Reviews. 2010;**32**:243-255

[6] United Nations Millennium Development Goals Report. 2006. Available from: https://www.un.org/zh/millenniumgoals/pdf/MDGReport2006

[7] Joy EJM, Ander EL, Young SD, Black CR, Watts MJ, Chilimba ADC, et al. Dietary mineral supplies in Africa. Physiologia Plantarum. 2014;**151**:208-229

[8] Hefferon KL. Can biofortified crops help attain food security? Current Molecular Biology Reports. 2016;**2**(4):180-185

[9] Gilani GS, Nasim A. Impact of foods nutritionally enhanced through bio-technology in alleviating malnutrition in developing countries. Journal of AOAC International. 2007;**90**(5):1440-1444

[10] Perez-Massot E, Banakar R, Gomez-Galera S, Zorrilla-Lopez U, Sanahuja G, Arjo G, et al. The contribution of transgenic plants to better health through improved nutrition: Opportunities and constraints. Genes and Nutrition. 2013;**8**(1):29-41

[11] Bouis HE. Economics of enhanced micronutrient density in food staples. Field Crops Research. 1999;**60**:165-173

[12] Nestel P, Bouis HE, Meenakshi JV, Pfeiffer W. Biofortification of staple food crops. Journal of Nutrition. 2006;**136**:1064-1067

[13] Pfeiffer WH, McClafferty B. HarvestPlus: Breeding crops for better nutrition. Crop Science. 2007;**47**:S88-S100

[14] Hirschi KD. Nutrient biofortification of food crops. Annual Review of Nutrition. 2009;**29**:401-421

[15] Bazuin S, Azadi H, Witlox F. Application of GM crops in sub-Saharan Africa: Lessons learned from green revolution. Biotechnology Advances. 2011;**29**:908-912

[16] Das JK, Kumar R, Salam RA, Bhutta ZA. Systematic review of zinc fortification trials. Annals of Nutrition and Metabolism. 2013;**62**(1):44-56

[17] White PJ, Broadley MR. Biofortifying crops with essential mineral elements. Trends in Plant Science. 2005;**10**:586-593

[18] Zhao FJ, McGrath SP. Biofortification and phytoremediation. Current Opinion in Plant Biology. 2009;**12**:373-380

[19] Belton PS, Taylor JRN. Sorghum and millets: Protein sources for Africa. Trends in Food Science and Technology. 2004;**15**:94-98

[20] O'Kennedy MM, Grootboom A, Shewry PR. Harnessing sorghum and millet biotechnology for food and health. Journal of Cereal Science. 2006;**44**:224-235

[21] Agricultural and Processed Food Products Export Development Authority (APEDA). Available from: https://apeda.gov.in/milletportal/Production.html [accessed January 02, 2023]

[22] Food and Agricultural Organization [FAO]. 2015. Available from: http://faostat3.fao.org/ [accessed January 02, 2023]

[23] Austin D. Fox-tail millets (Setaria: Poaceae): Abandoned food in two hemispheres. Economic Botany. 2006;**60**:143-158

[24] Vijayakumari J, MushtariBegum J, Begum S, Gokavi S. Sensory attributes of ethnic foods from fingermillet (*Eleusine coracana*). In: Proceeding of the National Seminar on Processing and Utilization of Millet for Nutrition Security: Recent Trends in Millet Processing and Utilization. Hisar: CCSHAV; 2003. pp. 7-12

[25] Baltensperger DD. Progress with proso, pearl and other millets. In: Janick J, Whipkey A, editors. Trends in New Crops and New Uses. Alexandria, VA: ASHS Press; 2002. pp. 100-103

[26] Kimata M, Negishi M. Geographical distribution of prosomillet *Panicummiliaceum* L. on iodo starch and phenol reactions; with special references to a Northern Propagation Route into Japanese Islands. Environmental Education Studies. Tokyo Gakugei University. 2002;**12**:15-22

[27] Sood S, Khulbe RK, Gupta AK, Agrawal PK, Upadhyaya HD, Bhatt JC. Barnyard millet–a potential food and feed crop of future. Plant Breeding. 2015;**2015**(134):135-147

[28] de Wet JMJ, Rao KEP, Mengesha MH, Brink DE. Diversity in kodomillet, Paspalumscrobiculatum. Economic Botany. 1983b, 1983b;**37**:159-163

[29] de Wet JMJ, Rao KEP, Brink DE. Systematics and domestication of Panicum sumatrense (Gramineae). Journal of Agricultural Traditional Botany Application. 1983a;**30**:159-168

[30] Yadav OP, Rai KN. Genetic improvement of pearlmillet in India. Agricultural Research. 2013;**2**:275-292

[31] Yadav OP, Rai KN, Bidinger FR, Gupta SK, Rajpurohit BS, Bhatnagar SK. Pearlmillet (Pennisetumglaucum) restorer lines for breeding dual-purpose hybrids adapted to arid environments. Indian Journal of Agricultural Sciences. 2012;**82**:922-927

[32] Agricultural Statistics, Government of India. PearlMillet. In: Agricultural Statistics at a Glance–2014, Government of India. New Delhi: Oxford University Press; 2014. pp. 85-86

[33] Nedumaran S, Bantilan MCS, Gupta SK, Irshad A, Davis JS. Potential Welfare Benefit of Millets Improvement Research at ICRISAT: Multi Country-Economic Surplus Model Approach, Socioeconomics Discussion Paper SeriesNumber15. Hyderabad: ICRISAT; 2014

[34] Hegde PS, Rajasekaran NS, Chandra TS. Effects of the antioxidant properties of millet species on oxidative stress and glycemic status in alloxan induced rats. Nutrition Research. 2005;**25**:1109-1120

[35] Saleh ASM, Zhang Q, Chen J, Shen Q. Milletgrains: Nutritional quality, processing, and potential health benefits. Comprehensive Reviews in Food Science and Food Safety. 2013;**12**:281-295

[36] Oberleas DP. Phytates. In: Strong FM, editor. Toxicants Occurring Naturally in Foods. Washington DC: National Academy of Sciences; 1973. pp. 363-371

[37] Gupta VP. Genetics of quality improvement. In: Gupta VP, Minnocha JL, editors. Trends in Genetic al Research on Pennisetums. Vol. 1980. Ludhiana: Punjab Agricultural University; 1980. pp. 291-294

[38] Kumar A, Chauhan BM. Effect of phytic acid on protein digestibility (in vitro) and HCl-extractability of minerals in pearlmillet sprouts. Cereal Chemistry. 1993;**70**:504-506

[39] Abdalla AA, ElTinay AH, Mohamed BE, Abdalla AH. Proximate composition, starch, phytate and mineral contents of 10 pearlmillet genotypes. Food Chemistry. 1998;**63**:243-246

[40] AbdelRahman SM, Babiker EE, ElTinay AH. Effect of fermentation on anti nutritional factors and HCl extractability of minerals of pearlmillet cultivars. Journal of Food Technology. 2005;**3**:516-522

[41] Shivaraj B, Pattabiraman TN. Natural plant enzyme inhibitors. Characterization of an unusual alpha-amylase/trypsin inhibitor from ragi (Eleusinecoracana Geartn.). Biochemical Journal. 1981;**193**:29-36

[42] Pattabiraman TN. Trypsin/chymo trypsin inhibitors from millets. Adv. Exp. Med. Biol. 1986;**199**:439-448

[43] Joshi BN, Sainani MN, Bastawade KB, Deshpande VV, Gupta VS, Ranjekar PK. Pearlmillet cysteine protease inhibitor. Evidence for the presence of two distincts it's responsible for anti-fungal and anti-feedant activities. European Journal of Biochemistry. 1999;**265**:556-563

[44] McDonough CM, Rooney LW, Serna-Saldivar SO. In: Kulp K, Ponte JG, editors. The Millets. In: Hand Book of Cereal Science and Technology (Food Science and Technology). New York, NY: Marcel Dekker Inc; 2000. pp. 177-202

[45] Hadimani NA, Muralikrishna G, Tharanathan RN, Malleshi NG. Varying in processing characteristics and kernel texture. Journal of Cereal Science. 2000;**33**:17-25

[46] Chandrasekara A, Shahidi F. Content of insoluble bound phenolics in millets and their contribution to anti-oxidant capacity. Journal of Agricultural and Food Chemistry. 2010;**58**:6706-6714

[47] Zhang LZ, Liu RH. Phenolic and carotenoid profiles and anti-proliferative activity of foxtail millet. Food Chemistry. 2015;**174**:495-501

[48] Bhatt D, Negi M, Sharma P, Saxena SC, Dobriyal AK, Arora S. Responses to drought induced oxidative stress in five finger millet varieties differing in their geographical distribution. Physiology and Molecular Biology of Plants. 2011;**17**:347-35356

[49] Devi PB, Vijayabharathi R, Sathyabama S, Malleshi NG, Priyadarisini VB. Health benefits of finger millet (*Eleusine coracana* L.) polyphenols and dietary fiber: A review. Journal of Food Science and Technology. 2014;**51**:1021-1040

[50] Chandrasekara A, Shahidi F. Inhibitory activities of soluble and bound millet seed phenolics on free radicals and reactive oxygen species. Journal

of Agricultural and Food Chemistry. 2011;**59**:428-436

[51] Kofuji K, Aoki A, Tsubaki K, Konishi M. Antioxidant activity of b-glucan. ISRN Pharm. 2012:125864. E pub 2012, Feb 19. DOI: 10.5402/2012/125864

[52] Sharma S, Saxena DC, Riar CS. Isolation of functional components b-glucan and g-amino butyric acid from raw and germinated barnyard millet (*Echinochloa frumentacea*) and their characterization. Plant Foods for Human Nutrition. 2016;**71**:231-238

[53] Dykes L, Rooney LW. Review: Sorghum and millet phenols and antioxidants. Journal of Cereal Science. 2006;**44**:236-251

[54] Dykes L, Rooney LW. Phenolic compound sin cereal grains and their health benefits. Cereal Foods World. 2007;**52**:105-111

[55] Kim JS, Hyun TK, Kim MJ. The inhibitory effects of ethanol extracts from sorghum, foxtail millet and proso millet on a-glucosidase and a-amylase activities. Food Chemistry. 2011;**124**:1647-1651

[56] Kunyanga CN, Imungi JK, Okoh MW, Biesalsk HK. Total phenolic content, anti oxidant and anti diabetic properties of methanolic extract of raw and traditionally processed Kenyan indigenous food ingredients. LWT - Food Science and Technology. 2012;**45**:269-276

[57] Muthamilarasan M, Prasad M. Advances in Setaria genomics for genetic improvement of cereals and bio energy grasses. Theoretical and Applied Genetics. 2015;**128**:1-14

[58] O'Kennedy MM, Crampton BG, OziasAkins P. Pearlmillet. In: Kole C, Hall TC, editors. Compendium of Transgenic Crop Plants: Transgenic Cereals and Forage Grasses. Vol. 1. Chichester: John Wiley & Sons; 2009. pp. 177-190

[59] Tsehaye Y, Berg T, Tsegaye B, Tanto T. Farmers' management of fingermillet (*Eleusine coracana* L.) diversity in Tigray, Ethiopia and implications for on-farm conservation. Biodiversity and Conservation. 2006;**15**:4289-4308

[60] Reddy INBL, Reddy DS, Narasu ML, Sivaramakrishnan S. Characterization of disease resistance gene homologues isolated from finger millet (*Eleusinecoracana* L. Gaertn). Molecular Breeding. 2011;**27**:315-328

[61] Bouis HE, Hotz C, McClafferty B, Meenakshi JV, Pfeiffer WH. Biofortification: A new tool to reduce micro nutrient malnutrition. Food and Nutrition Bulletin. 2011;**32**:S31-S40

[62] Welch RM, Graham RD. Breeding for micro nutrients in staple food crops from a human nutrition perspective. Journal of Experimental Botany. 2004;**55**:353-364

[63] Kennedy GL, Pedro MR, Seghieri C, Nantel G, Brouwer I. Dietary diversity score is a useful indicator of micronutrient intake in non-breastfeeding Filipino children. The Journal of Nutrition. 2007;**137**:472-477

[64] WHO. WHO/FAO (World Health Organization/Food and Agriculture Organization of the United Nations). Preparation and Use of Food-Based Dietary Guidelines. Nutrition Programme. Geneva: World Health Organization; 2003

[65] World Health Organization. Office of World Health Reporting. The World Health Report: Reducing Risks, Promoting Healthy Life: Overview.

Geneva: World Health Organization; 2002. ISBN 924 1562072. Available from: https://apps.who.int/iris/handle/10665/67454

[66] WHO Newsroom. Factsheet "Children: Reducing Mortality". 2019. Available from: https://www.who.int/newsroom/fact-sheets [Accessed on January 02, 2023]

[67] Fuge R, Johnson CC. Iodine and human health, the role of environmental geochemistry and diet, a review. Applied Geochemistry. 2015;**63**:282-302

[68] Taylor JRN, Emmambux MN. Gluten-free foods and beverages from millets. In: Arendt EK, DalBello F, editors. Gluten-Free Cereal Products and Beverages (Food Science and Technology). Cambridge, MA: Academic Press; 2008. pp. 119-148

[69] Muthamilarasan M, Dhaka A, Yadav R, Prasad M. Exploration of millet models for developing nutrient rich graminaceous crops. Plant Science. 2016;**242**:89-97

[70] Hotz C, McCafeerty B, Hawkes C, Ruel M, Babu S. From harvest to health: Challenges for developing biofortified staple foods and determining their impact on micronutrient status. Food and Nutrition Bulletin. 2007;**28**:271-279

[71] Reddy BVS, Ramesh S, Longvah T. Prospects of breeding for micronutrients and β-carotene-dense sorghums. International Sorghum Millets Newsletter. 2005;**46**:10-14

[72] Velu G, Rai KN, Muralidharan V, Kulkarni VN, Longvah T, Raveendran TS. Prospects of breeding biofortified pearl millet with high grain iron and zinc content. Plant Breeding. 2007;**126**:182-185

[73] Rai KN, Govindraj M, Rao AS. Genetic enhancement of grain iron and zinc content in pearl millet. Quality Assurance and Safety of Crops & Foods. 2012;**4**(3):119-125

[74] Shahane AA, Shivay YS. Agronomic biofortification of crops: Current research status and future needs. Indian Journal of Fertilisers. 2022;**18**(2):164-179

[75] Satyavathi CT, Ambawat S, Khandelwal V, Govindaraj M. Micronutrient Rice Pearl Millet for Nutritionally Secure India. CAZRI, Jodhpur, Rajasthan: ICAR-All India Co-ordinated Research Project on Pearl Millet; 2021

[76] Yadava DK, Choudhury PR, Hossain F, Kumar D. Biofortified Varieties: Sustainable Way to Alleviate Malnutrition. Third ed. ICAR: New Delhi; 2020. pp. 1-86

[77] Kumar AA, Reddy BVS, Ramaiah B, Sahrawat KL. Gene effects and heterosis for grain iron and zinc concentration in sorghum [*sorghum bicolour* (L.) Moench]. Field Crops Research. 2013;**146**:86-95

[78] Mayer AB, Latham MC, Duxbury JM, Hassan N, Frongillo EA. A food systems approach to increase dietary zinc intake in Bangladesh based on an analysis of diet, rice production and processing. In: Thompson B, Amoroso L, editors. Combating Micronutrient Deficiencies: Food-Based Approaches. Rome, Italy and Wallingford, UK: FAO and CABI; 2011. pp. 254-300

[79] Sahrawat KL, Rago TJ, Wani S, Pradhasaradhi G. Sulfur, boron, and zinc fertilization effects on grain and straw quality of maize and sorghum grown in semi-arid tropical region of India. Journal of Plant Nutrition. 2008;**2008**(31):1578-1584

[80] Kumar S, Pandey G. Biofortification of pulses and legumes

to enhance nutrition. Heliyon. 2020, 2020;**6**(3):03682

[81] Dhawi F, Datta R, Ramakrishna W. Mycorrhiza and PGPB modulate maize biomass, nutrient uptake and metabolic pathways in maize grown in mining impacted soil. Plant Physiology and Biochemistry. 2015;**97**:390-399

[82] Dhawi F, Datta R, Ramakrishna W. Mycorrhiza and heavy metal resistant bacteria enhance growth, nutrient uptake and alter metabolic profile of sorghum grown in marginal soil. Chemosphere. 2016;**157**:33-41. DOI: 10.1016/j.chemosphere.2016.04.112

[83] Patidar M, Mali AL. Effect of farmyard manure, fertility levels and biofertilizers on growth, yield and quality of sorghum (Sorghum bicolor). Indian Journal of Agronomy. 2004;**2**(49):117-120

[84] Aro A, Alfthan G, Varo P. Effects of supplementation of fertilizers on human selenium status in Finland. The Analyst. 1995;**120**:841-843

[85] Cakmak I, Kalayci M, Ekiz H, Braun HJ, Kilinc Y, Yilmaz A. Zinc deficiency as a practical problem in plant and human nutrition in Turkey: A NATO-science for stability project. Field Crops Research. 1999;**60**:175-188

[86] Jiang XM, Cao XY, Jiang JY, Ma T, James DW, Rakeman MA. Dynamics of environmental supplementation of iodine: Four years' experience in iodination of irrigation water in Hotien, Xinjiang, China. Archives of Environmental & Occupational Health. 1997;**52**(6):399-408

[87] Rengel Z, Batten GD, Crowley DE. Agronomic approaches for improving the micronutrient density in edible portions of field crops. Field Crops Research. 1999;**60**:27-40

[88] Smith SE, Read DJ. Mycorrhizal Symbiosis. 3rd ed. London, UK: Elsevier; 2007

[89] Hardarson G, Broughton WJ. Maximising the use of biological nitrogen fixation in agriculture. Annals of Botany. 2004;**93**(4):477

[90] Cavagnaro TR. The role of arbuscular mycorrhizas in improving plant zinc nutrition under low soil zinc concentrations: A review. Plant and Soil. 2008;**304**:315-325

[91] Joy EJM, Ander EL, Young SD, Black CR, Watts MJ, Chilimba ADC, et al. Global Food Security. 2017;**12**:8-14

[92] Finkelstein J, Mehta S, Udipi S, Ghugre P, Luna S, Wenger M, et al. A randomized trial of iron-biofortifiied pearl milletin school children in India. Journal of Nutrition. 2015;**145**:1576-1581

[93] Finley JW. Bioavailability of selenium from foods. Nutrition Reviews. 2006;**64**(3):146-151

[94] Prasad R, Shivay YS. Agronomic biofortification of plant foods with minerals, vitamins and metabolites with chemical fertilizers and liming. Journal of Plant Nutrition. 2020;**43**:1534-1554

[95] Shivay YS, Singh U, Prasad R, Kaur R. Agronomic interventions for micronutrient biofortification of pulses. Indian Journal of Agronomy. 2016;**61**:S161-S172

[96] Stein AJ, Nestel P, Meenakshi JV, Qaim M, Sachdev HPS, Bhutta ZA. Plant breeding to control zinc deficiency in India: How cost-effective is biofortification? Public Health Nutrition. 2006;**10**(5):492-501

[97] Ma G, Jin Y, Li Y, Zhai F, Kok FJ, Jacobson E, et al. Iron and zinc

deficiencies in China: What is feasible and cost-effective strategy? Public Health Nutrition. 2008;**11**(6):632-638

[98] Garcia Banuelos ML, Sida Arreola JP, Sanches E. Biofortification – Promising approach to increasing the content of iron and zinc in staple food crops. Journal of Elementary Science Education. 2014;**19**(3):865-888

[99] Velu G, Ortiz Monasterio I, Cakmak I, Hao Y, Singh RP. Biofortification strategies to increase grain zinc and iron concentrations in wheat. Journal of Cereal Sciene. 2014;**59**(3):365-372

[100] Cakmak I, Pfeiffer WH, McClafferty B. Biofortification of durum wheat with zinc and iron. Cereal Chemistry. 2010;**87**(1):10-20

[101] Bouis HE, Welch RM. Biofortification – A sustainable agricultural strategy for reducing micronutrient malnutrition in the global south. Crop Science. 2010;**50**:20-32

[102] HarvestPlus. Biofortification Prioritization Index (BPI). Washington, DC: International Food Policy Research Institute (IFPRI); 2013 Available from: http://www.harvestplus.org/content

[103] Shahane AA, Shivay YS. Soil health and its improvement through novel agronomic and innovative approaches. Fronitiers in Agronomy. 2021;**3**:680456

[104] Bhattacharyya R, Ghosh BN, Mishra PK, Mandal B. Soil degradation in India: Challenges and potential solutions. Sustainability. 2015;7:3528-3570

[105] Carvalho SMP, Vasconcelos MW. Producing more with less: Strategies and novel technologies for plant-based food biofortification. Food Research International. 2013;**54**:961-971

Chapter 6

Blast Disease of Millets: Present Status and Future Perspectives

T. Tharana Poonacha, C.H. Sai Bhavana, Farooqkhan, G.V. Ramesh, Netravati Gavayi, Prasanna S. Koti, K.B. Palanna, H. Rajashekara, G. Rajesh and I.K. Das

Abstract

Millet crops are affected by various biotic and abiotic stresses. Among biotic stresses, blast disease caused by *Pyricularia grisea* (finger, pearl and proso millets) and *Pyricularia setariae* (foxtail millet) is the most devastating and widespread disease that causes substantial grain and forage yield losses and is a key constraint to pearl millet, finger millet and foxtail millet production in most of finger millet growing areas, and recently, it is also reported in barnyard millet in few locations. This book chapter emphasizes mainly on occurrence, distribution, symptoms, yield loss, etiology, genetic diversity, mode of spread of the pathogen and survival and integrated disease management approaches for mitigating of disease. This information will be highly helpful for better understanding of the disease. Further, it will be useful to enhance production and productivity of millets and to reinforce the food and nutritional security in the developing countries of Asia and Africa continents where the millets are mainly grown as staple food crops.

Keywords: millet, blast, *Pyricularia*, management and novel strategies, blast disease resistances

1. Introduction

Millets are small-seeded cereal crops widely known for its nutraceutical importance as well as food and fodder. The most frequently cultivated millets are sorghum (*Sorghum bicolor* L. Moench), finger millet (*Eleusine coracana* (L.) Gaertn.), foxtail millet (*Setaria italica* (L.) P. Beauvois), pearl millet (*Cenchrus americanus* (L.) Morrone.), kodo millet (*Paspalum scrobiculatum* L.), little millet (*Panicum sumatrense* Roth ex Roem. and Schult.), proso millet (*Panicum miliaceum* L.), browntop millet (*Brachiaria ramosa* (L.) Stapf) and barnyard millet (*Echinochloa crusgalli* (L.) P. Beauvois). In India, millets are cultivated for both grain and fodder. Though millets are regarded as hardy crops, present day climate change has rendered most of them susceptible to many pathogens.

Many diseases of millets regularly appear in a severe form under different climatic conditions and cause considerable economic loss, while others appear sporadically in specific climatic situations and have less damaging effect to the crop. Fungal diseases are more than bacterial and viral diseases. Important fungal diseases of millets are grain mold, ergot, smut, anthracnose, downy mildew, blast, rust, charcoal rot, foot rot, banded sheath blight and sheath rot. The diseases infect different plant parts including root, stem, leaves, peduncle or grain and adversely affect yield and quality of the produce. These diseases assume different significances for seed production, certifications and marketing of millets.

Among all these diseases, blast is a serious disease of millets especially in finger millet, pearl millet and foxtail millet. It is caused by the fungi, *Pyricularia grisea* (finger, pearl and proso millets) and *Pyricularia setariae* (foxtail millet). In finger millet crop is affected at all stages of its growth, and the disease is diagnosed by the production of elliptical- or diamond-shaped lesions on the leaf, peduncle and fingers, depending on the stage of the crop. The most damaging stage is neck blast, followed by finger and leaf blast. In foxtail millet, barnyard millet, proso millet and pearl millet symptoms are confined to only leaf. Moderate temperature, high humidity, cloudy days with intermittent rainfall conditions are ideal for the quick disease spread. The disease occurs almost every year in most of the finger millet and foxtail millet growing areas during rainy season, and in other millets, the disease is confined to specific location. The yield loss varies depending on the time of onset of the disease, severity, plant variety and prevailing weather.

1.1 Blast

Blast caused by *Pyricularia* spp. is one of the serious threats and most destructive disease that occurs widely in major millet growing regions of world. It is the major production constraints under natural conditions especially in finger, pearl and foxtail millet cultivation causing considerable economic losses with varying degrees of damage. In India, the finger millet blast was first reported from Tanjore delta of Tamil Nadu [1]. While foxtail millet blast was recorded in 1917 by Nishikado from Japan [2], but in India, it was reported in 1919 from Tamil Nadu [1], which further has also been recorded from Maharashtra, Andhra Pradesh [3] and Uttarakhand [4]. Since 1970, blast disease in pearl millet has been prevalent in major growing states of India; increased incidence has been reported recently in most pearl millet growing states like Gujarat, Uttar Pradesh, Madhya Pradesh, Rajasthan, Delhi, Maharashtra and Karnataka [5]. The disease is prevalent in all the major millet growing areas and spreading to new location as well with emerging pathotypes showing varying intensities depending on the cultivar, favorable conditions and production techniques.

1.2 Etiology

Pyricularia grisea (Cooke.) Sacc. [Perfect stage: *Magnaporthe grisea* (Herbert) Barr] causing blast in finger and proso millet whereas *Pyricularia setariae* Y. Nisik. infects foxtail millet. Kulkarni and Patel [6] grouped *P. setariae* into four physiological races on the basis of physiological, cultural, morphological characters and pathogenic ability of the fungus. However, Gaikwad and D′ Souza [7] determined that the isolates of *P. setariae* that infect foxtail millet differ from those that infect rice, finger millet and pearl millet. In case of pearl millet, *Pyricularia grisea* is known to cause blast disease in pearl millet. However, recently Singh et al. [8] reported that the foliar blast of pearl millet in western arid Rajasthan, India, is caused by *Pyricularia pennisetigena.*

1.3 Diagnostic symptoms

Blast pathogen can infect all the stages of plant in both finger and foxtail millet, the young seedlings are more prone for the attack and showed burnt appearance in nursery under severe infection [9]. In finger millet, *P. grisea* attacks at different growth stages of the crop and leads to formation of typical symptoms like leaf blast, neck blast and finger blast while in case of foxtail millet, *P. setariae* attacks the leaf lamina producing leaf blast symptoms [10–13].

On leaf lamina, the pathogen produces typical symptoms of water-soaked, spindle- or diamond-shaped lesions which are initially surrounded by chlorotic halo. Typical leaf blast symptoms are the formation of elliptical- or diamond-shaped lesions containing grayish center with dark brown margins. Under severe infection, adjacent lesions enlarge and may coalesce to form large necrotic areas which gives the crop burnt appearance from far. The pathogen infects and develops lesions on the leaf, peduncle and finger depending on the stage of the crop. The most devastating stage of finger millet blast is neck blast, in which the pathogen targets the neck region, reducing the number and weight of grain per earhead and leading in earhead sterility [14]. In this, neck portion of 2–4 inches below the ear immediately turns initially brown and later to black, where olive gray fungal growth can be observed in the blackened portion under high humid climate. In finger blast where the pathogen attacks fingers, *i.e.,* attacks usually the apical portions running towards the base (**Figure 1**). Infection of finger blast results in shriveled and blackened seeds which makes unfit for seed purpose and human consumption because of loss of minerals and vitamins. Ramakrishnan [15] observed spindle-shaped dark brown leaf spots 1–3 mm in length with grayish center and brownish periphery on finger millet, rice and *Digitaria* spp. leaves.

Symptoms of foxtail millet leaves mainly developed as from a small water-soaked yellowish dot, which later turned circular to an oval spot with a grayish center surrounded by a brown margin. Spots measured an average 2–5 mm in diameter within 2–3 days. The spots then coalesce and resulted in drying of leaves. The disease starts with the lower leaves and extend to upper leaves. No symptoms were observed on neck of foxtail millet [16]. Sharma *et al.* [17] observed blast disease symptoms on

Figure 1.
Typical blast disease symptoms of millets a&b) leaf blast of finger millet, c&d) neck blast of finger millet, e & f) finger blast of finger millet, g&h) blast of foxtail millet, i) blast of barnyard millet and blast of pearl millet.

foxtail millet leaves as tiny circular spots with gray-colored centers measuring 3–5 mm in diameter surrounded by a brown margin and also observed high disease severity in dense plant stand with moist condition.

In barnyard millet, the symptoms appear on the young seedlings under the field conditions. The spots are spindle to circular shaped with varying sizes. Initially the spots showed yellowish margin with grayish center. Later, the centers turned ash colored. Fungus develops an olive-gray overgrowth at the center of the spots under humid conditions [10].

1.4 Mode of spread and survival

The blast fungus, *Pyricularia,* can invade the host either by piercing the epidermal cells directly or through stomatal opening. Pyriform air borne conidia serves as both primary and secondary source of inoculum. The pathogen survives on infected host species or on weed hosts.

1.5 Host range

Finger millet, proso millet, foxtail millet, pearl millet, rice and wheat, etc., are infected with the pathogen. Nagaraja *et al.* [18] described that *P. grisea* isolated from finger millet possess the potential to infect rice crop but not *vice-versa*. Likewise, *P. setariae* isolated from foxtail millet shows the ability to infect finger millet, pearl millet, wheat and *Dactyloctaenium aegyptium* [19].

Mackill and Bonman [20] proposed that diverse weed hosts growing adjacent to the cultivated plants could serve as possible sources of inoculum for the disease, providing the fungus with an alternate method of survival. Despite the fact that blast infects a wide variety of sympatric flora Hamer *et al.* [21] and Valent *et al.* [22] determined that *M. grisea* populations are strongly confined by host range. Under experimental conditions, inoculations of rice with isolates of from weeds resulted in successful [20] and unsuccessful [23] cross-inoculations. Viji *et al.* [24] reported that in the laboratory, ten isolates of *M. grisea* from rice did not infect finger millet and vice versa, confirming that the *M. grisea* populations infecting rice and fibger millet in India were distinct. Similar results were reported by Kato et al. [25] and Todman *et al*. [26], who found that *Magnaporthe* isolates from *Eleusine coracana* failed to incite disease on rice and vice versa. Contradictory results were reported by Kumar and Singh [27] which could be attributed to prevailing environmental conditions during the experiments and the soil's nutritional level [28, 29].

Pennisetum is a diverse genus with over 100 species [30]. Susceptibility of all the species of *Pennisetum* to *Magnaporthe grisea* infection is not yet clear. The available information indicates that the pathogen infects principally *Pennisetum glaucum*, *P. macroforum, P. squamulatum*, *P. pedicellatum* [31], *P. ciliare* [32], *P. purpureum* [33]. Other graminaceous hosts such as *Agrostis palustris*, *Brachiaria mutica*, *Eleusine indica, Cyperus rotundus*, *Eragrostis* sp., *Panicum miliaceum* serve as collateral hosts for the pathogen [34].

1.6 Epidemiology

The crop is susceptible to the blast disease during all stages of its growth, *i.e.,* seedling (vegetative) to grain formation (reproductive) stage. Especially, young seedlings more prone to the blast both in the nursery and field conditions with favorable weather [35]. Moderate temperature (25–30°C) with high relative humidity (>90%)

and cloudy days coupled with intermittent rainfall creating continuous leaf wetness for more than 10 hr. are congenial for rapid development and spread of the disease. Continuous rains during heading lead to the occurrence of finger blast, resulting in massive production losses in both finger and foxtail millet. Also, high nitrogen fertilizer application is reported to increase blast disease [36].

1.7 Economic importance

Finger millet blast is economically one of the most important diseases, while blast of proso and foxtail millet are relatively of minor occurrence. The disease occurs almost every year in finger millet during rainy season, and losses vary with the time of onset of the disease, severity, cultivar and climatic conditions. During late 1970s to 80s, 1% incidence of finger and neck blast by *M. grisea* resulted in a corresponding enhancement of yield losses by 0.32 and 0.084% for neck and finger blast, respectively. Grain yield losses in finger millet, on the other hand, ranged from 6.75 to 87.5% [37]. In its severe form, foxtail millet blast can lead up to 30–40% loss of economic yield [10] while mean yield loss of finger millet blast ranged from 28 to 36% and may go up to 90% in endemic areas with frequent disease [38]. In pearl millet also the blast disease causes considerable yield loss under favorable environmental conditions (**Table 1**).

1.8 Disease cycle

The pathogen harbors in glumes, straw as well as on some graminaceous weeds. The blast pathogen is seed-borne with presence of inoculum in the pericarp and endosperm [19]. Blast fungal life cycle is complex due to its nature of disease which shows sensitivity to the weather conditions, survival and spread inoculum in different ways. During off-season, i.e., in the absence main host, it survives on the graminaceous weeds as collateral hosts who provides the primary inoculum for onset of infection. Further, the fungus spreads mainly by airborne conidia and occasionally through seeds.

1.9 Characterization of the pathogen

For proper diagnosis of the disease, the understanding of the pathogenic characteristics is needed as much of knowing symptomatology and disease cycle. Blast

Crop	Yield loss	Place	Ref
Pearl millet	13.3–14.9	India	[39]
Finger millet	42%	Ethiopia	[40]
	50 to 100%	TN, India	[41]
	35.78%	Gujarat, India	[42]
	22.57 to 56.67%	Karnataka, India	[43]
Foxtail millet	30–40%	India	[10]
	60%	India	[44]
	40%	India	[45]

Table 1.
Yield loss caused by blast disease in different crops.

caused by the *Pyricularia* spp. is identified based on its above-described symptoms in the field while *in vitro,* pathogen characterized based on cultural-morphological and molecular attributes. Morphological characterization includes studying mycelial features on agar plates, viz. appearance, color and amount of melanin pigment produced as well as the microscopic conidial characters. Molecular characterization of pathogen includes amplification of targeted genomic regions with fungal universal primers (*ITS)* as well as secondary barcoding regions such as *beta tubulin*, *TEF* and *LSU* and also by studying the DNA polymorphism using various molecular markers [46].

1.10 Morphology

M. grisea is a haploid, filamentous *Ascomycete* with morphological traits such as three-septate fusiform ascospores and black nonstromatic perithecia (ascocarp) with long hairy necks. The asexual stage *Pyricularia grisea* produces Conidia which are pyriform to obclavate, narrowed towards tip, rounded at the base, solitary, 2-septate, hyaline to pale brown, with a distinct basal hilum, sometimes with marginal frill. Studies on growth of *P. penniseti* on different media by Lukose *et al.* [47] indicated medium containing pearl millet leaf extract enriched with dextrose supported maximum growth of the pathogen. Light brown submerged growth was observed in potato dextrose agar medium, while pearl millet leaf extract medium showed grayish white superficial growth. Konda [48] tested effect of ten different solid media on growth of *P. setariae* and reported that maximum radial growth was observed in oat meal agar, PDA and malt extract agar followed by host leaf decoction +2 per cent sucrose agar medium. *M. oryzae* isolates were producing dull white to grayish-black colonies with regular margins. Conidia were pyriform, hyaline to pale olive and measured 16–23 x 4–7 µm in size [49].

Based on cultural and conidial variation, Viji *et al.* [24] differentiated *Pyricularia* isolates from different hosts. Sonah *et al.* [50] investigated that the cultural morphological variability of *M. grisea* isolates isolated from rice and other hosts and discovered that isolates with fast vegetative growth have gray-green or gray-white producing more spores than those with slower vegetative growth (submerged or subdued growth patterns). Isolates from non-rice hosts also have aberrant spore morphology, with longer, cylindrical and obpyriform spores. They also noticed a fair to good diversity in cultural and conidial characteristics among 17 field isolates of pearl millet [51].

1.11 Genetic diversity

Lot of information is available on variation among isolates of *Pyricularia* infecting various hosts like cereals and grasses. The pathogen is well studied in rice and is unfathomed in millets. The *Magnaporthe grisea* repeat sequence MGR586 was commonly used for studying population genetics of rice. Similarly fungal repetitive DNA or transposable elements are widely used for the purpose. The molecular level studies indicate the presence of variations among the isolates within or across the hosts. Several researchers have used molecular markers like RFLP [52] and SSR [53] and grouped the Indian isolates of *M. grisea* into two distinct populations—one finger millet group and other foxtail and rice group. Shivakantkumar et al. [51]. reported significant genetic variation among 17 *M. grisea* isolates infecting pearl millet with ITS, inter simple sequence repeats (ISSR) and simple sequence repeats (SSR) markers.

1.12 Mating type

Sexual reproduction is known to be a significant source of genetic variation in many fungi. Sexual compatibility in *M. grisea* is determined by the presence of two alleles (idiomorphs) at a single mating-type locus designated MAT1.1 and MAT1.2. MAT1.1 and MAT1.2 of *M. grisea* have been cloned and sequenced using a genomic subtraction strategy. The perfect stage of *P. grisea* was first described by Hebert [54] in crosses between isolates from cereals and wild grasses. Since then, efforts have been made to produce perithecia successfully on artificial media under controlled conditions using hermaphroditic tester isolates from finger millet and rice. Although both mating types have been found in the same field at the same time. However, it has not yet been possible to observe the perfect state in nature. The mating type assay *Magnaporthe* population infecting millets revealed that the mating-types, male fertile, female fertile and hermaphrodite nature of fertility existed finger, foxtail and branyard millet in the country. Which indicates the possibility of sexual recombination in field level and which may lead to high variability in pathogenicity and diversity in *Magnaporthe* population adapted to millets in India. All the tested isolates of pearl millet showed unknown fertility in PCR assay with MAT primers. It indicates fertility of pearl millet isolates has to be confirmed using range of tester isolates [55].

2. Management strategy

The integrated management strategies as well as novel approaches need to be adopted for the management of blast disease in millets. The integrated disease management strategies (**Figure 2**) for mitigating of millet blast are as follows.

2.1 Cultural practices

Several agricultural practices such as timely sowing, maintaining optimum plant populations and spacing, timely weeding, balanced use of fertilizers, crop rotation, deep plowing during summer season, removal of crop residues from the field, cleaning of field bunds after crop season, uprooting the diseased plant from the field and burning, regulating irrigation water from entering into other field, *etc.*, will help in reducing chances of disease occurrence.

2.1.1 Adjustment of date of sowing

The choice of sowing date in relation to crop disease has one principal aim to reduce to a minimum the period over which infective agent meets the susceptible stage of the host. Early sowing reduces blast severity.

2.1.2 Optimization of plant population

Due to high plant density, humidity in the field is always high, temperature is low, there is lack of aeration, and pathogen grows rapidly. Maintaining optimal plant population in the field to reduce the relative humidity build up in the field help in reducing disease severity.

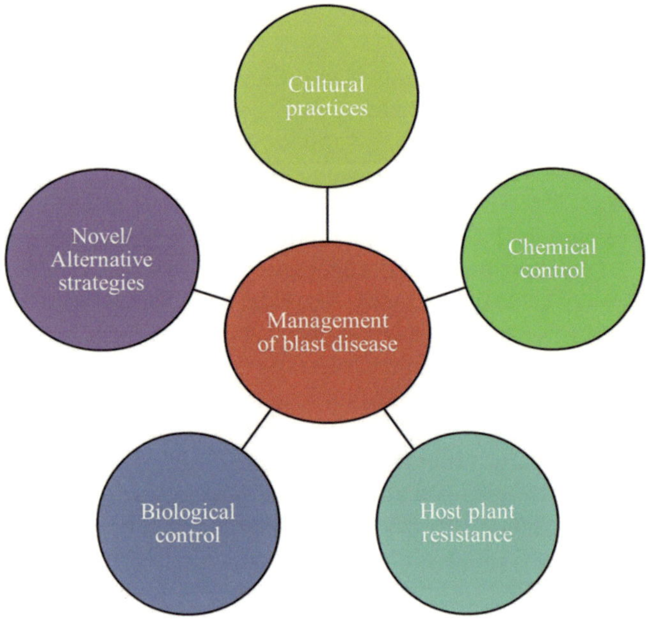

Figure 2.
Integrated disease management strategies of millet blast.

2.1.3 Use of disease-free seeds

Use of pathogen free seeds is a pre-requisite of eco-friendly control of plant disease because numerous plant pathogens are transmitted to the field *via* infected or contaminated seeds and seedlings.

2.1.4 Sanitation

Field sanitation is another important measure through cultural practices for preventing spread of plant disease and their management. Plant and plant parts are some of the best reservoirs of disease organisms. The inoculum present on few plants in the field may multiply on the plant and in due course of time may appear to cause epidemic in next season.

2.1.5 Eradication of alternate and collateral hosts

Plant pathogens usually have a wide host range and they used weeds, wild host plants or self-sown host plants as means of their active survival during absence of main crop. Therefore, keeping the field free from additional host of pathogen is a major sanitary cultural practice. Their timely removal helps to control blast disease.

2.1.6 Fertilizer management

A properly nourished plant is able to withstand or tolerate the attack of pathogens much better than a plant that has either nutrient deficiencies or excesses. A nutrient-deficient plant will be stressed and therefore more prone to disease attack. Excessive fertilizer applications can also cause plants to be more susceptible to disease. Nutrient can affect the relationship between crop and pathogen in many ways. Regulating the amount of nitrogenous fertilizer reduces incidence of blast and other diseases.

2.2 Host plant resistance

Exploiting host resistance to control disease is not only economical but also a practical necessity in a low value crop like millets where there is a limitation for any additional cash inputs such as fungicides, etc. Development of resistance varieties is the best way of combating the disease, which is primarily grown by resource-poor and marginal farmers. Disease resistant varieties identified and released for the different millets growing areas of India are tabulated in **Table 2**.

2.3 Biological control

Biological control is an alternative to synthetic chemical pesticides and having several benefits to human beings and ecosystem; they can ensure the protection of plants against biotic and abiotic stresses, production of good quality grains, improve soil fertility, sustainable and safety of environment. The demand for development and application of indigenous bioinoculant products has increased among researchers because of their role in plant growth promotion and crop protection in sustainable farming systems and also for their economic value. Bio-control agents especially strains of *Trichoderma* and *Pseudomonas* are useful for seed and soil borne diseases of millets.

In finger millet seed treatment with *P. fluorescens* @ 6 g/Kg seed and spray *P. fluorescens* formulations at 2 g/lit of water. First spray immediately after noticing the symptom. Second and Third sprays at flowering stage at 15 days interval was found effective for blast disease management.

2.4 Chemical control

Chemicals are not normally used for disease management in millet, because of involvement of high cost of chemical and labor. However, sometimes its use in combination with resistant cultivar becomes necessary. Fungicides are mostly used either as seed treatment or foliar spray. However, combination of them gives better management.

S.No.	Crop	Disease resistant/moderately resistant varieties	Reference
1	Pearl millet	BHB-1202, (MH 1831), Central Pearl Millet Hybrid RHB 223, JKBH 1008, MPMH 21, HHB 272, PB 1852, DHBH 1397, PROAGRO 9450, XMT 1497, Bio 8145, 86 M84 (MH 1890), KBH 108, 86 M88 (MH 1816) and ABV 04 (MP 552)	[56]
2	Finger millet	GPU 26, GPU 45, Chilika (OEB 10), VL 315, GPU 48, PRM 1, Bharathi (VR 762), Srichaitanya, KMR 301, KOPN 235, OEB 526, OEB 532, PPR 2700 (Vakula), VL 352, GNN - 6, GN - 5, VL Mandua - 348, KMR 340, Dapoli – 2 (SCN – 6), CO 15	[56]
3	Foxtail millet	RAU (Rajendra Kauni 1–2) and SiA 3085	[56]

Table 2.
Blast disease resistant varieties identified and released for the different growing areas of India (2000–2018).

Seed treatment with carbendazim @ 1 gm/Kg of seed. Spray any one of the fungicides *viz.*, Carbendazim (0.2%) or Iprobenphos (IBP) (0.1%) or premixture fungicide (Carbendazim+Mancozeb) (0.1%), Ediphenphos (0.1%) or propiconazole (0.1%) or Tricyclazole (0.1%). First spray immediately after noticing the symptoms. Need-based second and third sprays at flowering stage at 15 days interval to control neck and finger infection in finger millet. Similarly in rice several fungicides like Tricyclazole have been extensively tested and recommended [57], Probenazole [58], Isoprothiolane [59], Azoxystrobin, *etc.,* are recommend to control blast disease. Many fungicides are used against blast disease, including benomyl, iprobenfos, pyroquilon, felimzone, diclocymet, carpropamid and metominostrobin [60].

3. Future perspectives

With the everyday increasing population of world demands not only the food security but also the nutritional security which combined to form agricultural sustainability. Updated reports show that the agricultural production needs to be increased 50% by 2050 to meet the growing food and nutritional demand [61]. Nutritional security is as much important as food security for better quality of life which can be fulfilled by the cultivation of millets on large scale. Erstwhile commercial crops like rice, wheat, sugarcane, maize, etc., have been given more importance owing to their wide distribution and acceptance as daily food. In recent times, millets especially small millets gaining huge attention with growing health concerns which are fulfilling by their nutraceutical properties. This results in increasing area of cultivation, and further pathogen gets better opportunity for survival and spread; changes in virulence spectrum; emergence of new pathotypes/races, *etc.,* which intern results in breakdown of resistance as well as expansion of host rage, *etc.,* may lead to drastic reduction in production and productivity.

Therefore, systemic research in small millets on many aspects like breeding of new varieties, sequencing of genome, etc., are still underway has to be initiated on priority. With the advent of advanced genomic approaches like next generation sequencing (NGS), genome editing techniques, etc., made identification, cloning and transfer of resistant (R) genes easy. Using of such approaches in millets aid in better understanding of the crops and pave way for possible manipulation of crop genome to generate disease resistant crops which is an eco-friendly perspective. This will make the millets possibly the eco-friendly alternative for nutraceutical supplement, cost effective due to no use of pesticides and farmer friendly. Also, surveillance of the established diseases and regular monitoring of new diseases aid in achieving the food and nutritional security.

4. Novel/alternative strategies for enhancing blast disease resistances in millets

Millets are known to show exceptional tolerance to both biotic and abiotic stress in comparison with the popularly grown cereals. The stress resilience is due to various morpho physiological, biochemical and molecular mechanisms. Yet, in the recent era of drastic climate change, there is an increasing report of occurrence of blast disease in millets. The scenario of changing disease severity emphasizes the need to explore into novel, alternative strategies for disease management (**Figure 3**).

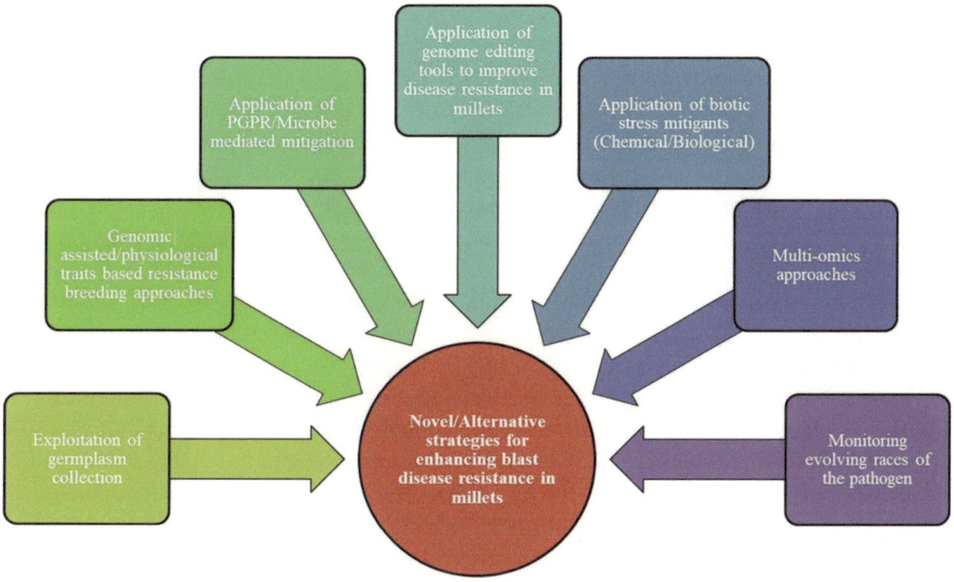

Figure 3.
Novel/alternative strategies for enhancing blast disease resistances in millets.

4.1 Exploitation of germplasm collection

Germplasm collections aid in the preservation of genetic resources for use in agricultural research and crop improvement programs around the world. Various millet germplasm has been preserved in various national and international gene banks. International Crop Research Institute for Semi-Arid Tropics (ICRISAT), Patancheru, in collaboration with different national and international organizations has enormous collections of accessions of millets with various traits of interest from across the world. India holds largest collection of millets at ICAR-National Bureau of Plant Genetic Resources, New Delhi, followed by All India Coordinated Small Millets Improvement Project (AICSMIP) at Bangalore and IIMR, Hyderabad. Germplasm could be a fantastic low-cost resource for allele mining and genotypic variant identification for blast disease resistance and other essential agronomic features. Mapping with genetic markers, identifying trait-specific germplasm and identifying candidate genes have a wide range of applications in millet improvement programmes. Lack of genetic resources hampers improvement of millets. Identifying and use new genes for host plant resistance in order to generate cultivars resistant to biotic stresses is critical.

4.2 Genomic assisted/physiological traits-based resistance breeding approaches

How plants adapt to external conditions determines plants functional trait (morphological, phenological, physiological and nutritional). Variation in these features results in one genotype being superior to others. Automation, imaging and software solutions have opened the way for many high-throughput phenotyping research in recent years. To conduct genetic studies or marker-aided breeding in any crop, DNA-based molecular markers, genetic linkage maps and sequence information are required genomic resources. Early diversity studies utilizing first generation molecular markers (RAPD, RFLP and ISSR) revealed limited sequence diversity in most millet populations [62], limiting their use in analyzing genetic variations in millet populations. Germplasm collections are used to select ideal mating parents for hybrid breeding, study population structure and analyze QTLs. On the other hand, adaptation of millets to diverse climate makes it impossible to rule out the possibility of the existence of considerable genetic diversity among the millets germplasms. The creation of relevant SNPs, as well as the mapping and tagging of QTLs related with blast resistance, would aid in the cloning of important disease resistance genes and the generation of resistant cultivars through a marker-assisted breeding programme [63].

4.3 Application of PGPR/microbe-mediated mitigation

Plant growth promoting bacteria (PGPB) promote growth and regulate plant development via cell proliferation and elongation, and development of lateral and adventitious roots by IAA production and ACC deamination. The various PGPB studies in millets include *Pseudomonas* sp., *Florescent pseudomonads, Enterobacter* sp. *PR14, Sphingomonas faenimutants, Acinetobacter calcoaceticus* and *Bacillus amyloliquefaciens* EPP90. These bacteria help in mitigating the effects of various abiotic stresses by an increased phosphate solubilization and antioxidant activity of enzymes and accumulation of osmo-protectants and a decreased lipid peroxidation [64]. Isolation and identification of such PGPR/ microbes with antimicrobial properties against blast disease would be the breakthrough in eco-friendly management of the blast disease in millets.

4.4 Application of biotic stress mitigants (chemical/biological)

Rice blast disease, caused by *Magnaporthe oryzae*, is one of the most devastating diseases worldwide. Many excellent chemicals for this particular disease viz., blasticidin S, kasugamycin, iprobenphos (IBP), edifenphos (EDDP), isoprothiolane, ferimzone and metominostrobin have been developed. Present reports of resistance towards fungicides prioritize now as the high time to focus our interest in non-fungicidal disease controlling agents since they are supposedly specific to target organisms along with least resistance development problems. Actually, two groups of non-fungicidal rice blast chemicals are currently on the market; melanin biosynthesis inhibitors (e.g., fthalide, tricyclazole, pyroquilon, carpropamid, diclocymet and fenoxanil) and the so-called priming effectors or plant defense activators (probenazole, acibenzolar-S-methyl and tiadinil) which induce host resistance against the pathogen's attack [65]. Helvolic acid, a terpenoid molecule extracted from the yeast *Pichia guilliermondii*. Isolated from the medicinal plant *Paris polyphylla*, was found to strongly inhibit *M. oryzae* spore germination. Cryptocin, isolated from the fungal endophyte *Cryptosporiopsis quercina*, which colonizes the inner stem bark tissue of *Tripterygium wilfordii*, similarly affects *M. oryzae* [66]. In comparison with the synthetic fungicides now in use, these plant extracts are non-hazardous, ecologically safer, locally available, renewable and easily accessible when used to manage rice blast disease.

4.5 Application of genome editing tools to improve disease resistance in millets

Of late development of cultivars employing genetic engineering technologies is gaining importance in plant biology and stress physiology. Understanding the mechanisms that regulate gene expression and the ability to transfer essential genes from other organisms into plants will broaden the ways in which plants can be utilized. In the near future, the employment of innovative approaches combining physiological, biochemical, molecular and genetic techniques could yield great results. Considering the geographical area of cultivation, available resources and the expertise of native researchers, millet genome editing is expected to progress at a slow pace. Due to limited expertise and infrastructure among millet research labs in developing countries the reach of modern tools like genome editing is delayed. Among the available genome editing tools, the CRISPR/Cas system may play a key role for genome editing in millets due to it user friendly and cost-effective construct design [67].

4.6 Multi-omics approaches

Most of the present information regarding the genome of millets is obtained through comparative genomics application with rice. With the availability of *Eleusine coracana*'s high-quality genome sequence, it will be possible to locate new selection targets and utilize genomic selection and prediction approaches. This will expedite the breeding process and will allow simultaneous selection for yield, quality and disease resistance. Up to date only draft genome sequences are released for finger millet [68], pearl millet [69] and proso millet [70]. Fully annotated genome sequences are not yet available for these millets. Not even a draft genome sequence is reported for other millets (little millet, barnyard millet and kodo millet) up to now. The cross genera transferability of the reported DNA markers demonstrates their usability in understanding phylogenetics. However, molecular breeding efforts utilizing omics tools and translational research are lagging in most of the millet crops [71]. Using transgenic technology or molecular-assisted breeding, the possible candidate genes found through various transcriptome and proteomic investigations could be used to generate cultivars with better adaptability to endure harsh climatic conditions.

4.7 Monitoring evolving races of the pathogen

In several Indian states, host-directed evolution of pathogenic variation has resulted in severe disease. As a result, understanding pathogen diversity, mode of action and genetics of host plant resistance is critical for developing successful crop improvement methods against biotic stressors. Despite significant progress in managing various fungal diseases in millet, there is still much room for improvement in developing disease management strategies, with a primary focus on virulence monitoring, identification and characterization of newer isolates and development of resistant hybrids/cultivars for commercial cultivation [72].

It is summarized that the blast disease poses a significant challenge to the millets production, now and in the future as evidenced by increase in disease incidence on pearl millet, finger millet and foxtail millet as well as the blast incidence is also observed on other millets viz., barnyard millet and brown top millet at few locations in moderate to severe form. In this chapter, attempts have been made to briefly summarize the key aspects of blast disease caused by *Magnaporthe* spp. and novel strategies for enhancing Blast disease resistances in millets. This chapter serves as a reference point for pathologists and breeders accompanied in field study in non-exhaustive manner to comprehend the complexity of diseases and to contemplate them in a more holistic manner.

Author details

T. Tharana Poonacha[1], C.H. Sai Bhavana[1], Farooqkhan[1], G.V. Ramesh[2], Netravati Gavayi[1], Prasanna S. Koti[3], K.B. Palanna[1*], H. Rajashekara[4], G. Rajesh[5] and I.K. Das[5]

1 ICAR-AICRP on Small Millets, Project Coordinating Unit, UAS, GKVK, Bengaluru, Karnataka, India

2 Department of Plant Pathology, Punjab Agricultural University, Ludhiana, Punjab, India

3 Department of Plant Biotechnology, UAS, GKVK, Bengaluru, Karnataka, India

4 Vivekananda Parvatiya Krishi Anusandhana Sansthan, Almora, Uttarakhand, India

5 ICAR-Indian Institute of Millets Research, Hyderabad, Telangana, India

*Address all correspondence to: kbpalanna@gmail.com

References

[1] McRae W. Detailed Administration Report of the Government Mycologist for the Year 1919-20. India: Madras Agric Dept; 1920

[2] Nishikado Y. Studies on the Rice blast fungus, (I). Berichte des Ohara Instituts für Landwirtschaftliche Forschungen. 1917;**1**(2):171-218

[3] Sinha AP, Upadhyay JP. Millet Ke. Rog. Directorate of Publication. Pantnagar, Uttarakhand: G.B. Pant University of Agriculture and Technology; 1997. p. 180

[4] Kumar B. Diseases of small millets in Uttarakhand and their management. In: Innovative approaches in plant disease management-crop diseases and their management. Singh KP, Prajapati CR, Gupta AK, editors. Germany: LAP Lambert Academic Publishing; 2013. pp. 257-287

[5] Chandra NS, Srivastava RK, Lavanya SN, Prakash G, Bishnoi HR, Kadvani DL, et al. Magnaporthe Blast of Pearl Millet in India. Jodhpur: All India Coordinated Research Project on Pearl Millet (Indian Council of Agricultural Research); 2017. p. 51

[6] Kulkarni NB, Patel MK. Study of the effect of nutrition and temperature on the size of spores in *Pyricularia setariae*. Indian Phytopath. 1956;**9**:33-38

[7] Gaikwad SP, D'Souza TF. A comparative study on Pyricularia spp. Journal of Maharashtra Agricultural Universities (India). 1987;**12**(1):134-135

[8] Singh SK, Solanki RK, Kakani RK. Pearl millet blast disease caused by Pyricularia pennisetigena in western arid Rajasthan, India. Current Science. 2020;**119**(10):00113891

[9] Parthasarathy S, Thiribhuvanamala G, Prabakar K. Diseases of field crops and their management. CRC Press; 2020

[10] Nagaraja A, Kumar J, Jain AK, Narasimhadu Y, Raghuchander T, Kumar B, et al. Compendium of small millets diseases. In: Project Coordinator Cell. Bengaluru: All India Coordinated Small Millets Improvement Project, UAS, GKVK Campus; 2007. p. 80

[11] Rajashekara H, Mishra KK, Pattanayak A. Blast disease in hill crops: Symptoms, identification and management. In: Stanley J, Mishra KK, Subbanna ARNS, Rajashekara H, Pattanayak A, editors. Integrated Pest Management in Major Crops. Almora: ICAR-VPKAS; 2019. pp. 38-48

[12] Rajashekara H, Jeevan B, Mishra KK, Subbanna ARNS, Kant L. Blast disease: Historical importance, distribution, and host infectivity across cereal crops. In: Nayaka SC, Hosahatti R, Prakash G, Satyavathi CT, Sharma R, editors. Blast Disease of Cereal Crops. Fungal Biology. Cham: Springer; 2021. pp. 1-13

[13] Palanna KB, Rajashekara H, Ramesh GV, Malikarjuna B, Praveen B, Raveendra HR, et al. Current scenario and integrated approaches for Management of Finger Millet Blast (*Magnaporthe grisea*). In: Nayaka SC, Hosahatti R, Prakash G, Satyavathi CT, Sharma R, editors. Blast Disease of Cereal Crops. Fungal Biology. Cham: Springer; 2021. pp. 27-49

[14] Rath GC, Mishra D. Nature of losses due to neck blast infection in ragi. Science & Culture. 1975;**41**(7):322-323

[15] Ramakrishnan KV. Studies on the morphology, physiology and parasitism

of the genus Pyricularia in Madras. Proceedings of the Indian Academy of Sciences-Section B. 1948;**27**(6):174-193

[16] Palaniswami A, Govindaswamy CV, Vidhyasekaran P. Studies on blast disease of Tenai. Madras Agricultural Journal. 1970;**57**:686-693

[17] Sharma R, Girish AG, Upadhyaya HD, Humayun P, Babu TK, Rao VP, et al. Identification of blast resistance in a core collection of foxtail millet germplasm. Plant Disease. 2014;**98**(4):519-524

[18] Nagaraja A, Das IK, Tonapi VA. Diseases of Millets- a Ready Reckoner. Rajendranagar, Hyderabad 500030, Telangana: Indian Institute of Millets Research; 2016. p. 67

[19] Viswanath S, Seetharam A. Diseases of small millets and their management in India. In: Seetharam A, Riley KW, Harinarayana G, editors. Small Millets in Global Agriculture. New Delhi: Oxford & IBH Publishing Co. Pvt. Ltd; 1989. pp. 237-253

[20] Mackill AO, Bonman JM. New hosts of Pyricularia oryzae. Plant Disease. 1986;**70**(2):125-127

[21] Hamer JE, Farrall L, Orbach MJ, Valent B, Chumley FG. Host species-specific conservation of a family of repeated DNA sequences in the genome of a fungal plant pathogen. Proceedings of the National Academy of Sciences. 1989;**86**(24):9981-9985

[22] Valent B, Crawford MS, Weaver CG, Chumley FG. Genetic studies of fertility and pathogenicity in Magnaporthe grisea (Pyricularia oryzae). Iowa State Journal of Research. 1986;**60**(4):569-594

[23] Prabhu AS, Filippi MC, Castro N. Pathogenic variation among isolates of Pyricularia oryzae affecting rice, wheat, and grasses in Brazil. International Journal of Pest Management. 1992;**38**(4):367-371

[24] Viji G, Gnanamanickam SS, Levy M. DNA polymorphisms of isolates of Magnaporthe grisea from India that are pathogenic to finger millet and rice. Mycological Research. 2000;**104**(2):161-167

[25] Kato H, Yamaguchi T, Nishihara N. Seed transmission, pathogenicity and control of ragi blast fungus and susceptibility of ragi to Pyricularia spp. from grasses, cereals and mioga. Japanese. Journal of Phytopathology. 1977;**43**(4):392-401

[26] Todman AK, Pawar DR, Joshi MH. Host reactions to finger millet blast (Pyricularia grisea Sacc.). Mysore journal of. Agricultural Sciences. 1994;**28**:45-46

[27] Kumar AR, Singh RA. Differential response of Pyricularia grisea isolates from rice, finger millet and pearl millet to media, temperature, pH and light. Indian Journal of Mycology and Plant Pathology (India). 1995;**25**(3):238-242

[28] Asuyama H. Morphology, taxonomy, host range, and life cycle of Piricularia oryzae. The Rice Blast Disease. 1965. p. 9-12

[29] Ou SH. Rice diseases second edition. Common Wealth Mycological Institute Kew England. 1985;**380**

[30] Ojo OA, Ojo AB, Morayo B, Iyobhebhe M, Elebiyo TC, Evbuomwan IO, et al. Phytochemical properties and pharmacological activities of the genus Pennisetum: A review. Scientific African. 2022;**16**:e01132

[31] Saikai UN, Puzari KC, Dutta PK. *Pennisetum pedicellatum*, a new host of

Pyricularia grisea. Journal of Mycology Plant Pathology. 1983;**12**:237

[32] Perrott RF, Chakraborty. *Pyricularia grisea* causes blight of buffel grass (*Cenchrus ciliaris*) in Queensland, Australia. Trop Grasslands. 1999;**33**:201-206

[33] Buckley TA, Allen BF. Notes on current investigations, April to June, 1951. Cacao. Malayan Agricultural Journal. 1951;**34**:134-135

[34] Singh S, Sharma R, Chandra Nayaka S, Tara Satyavathi C, Raj C. Understanding pearl millet blast caused by Magnaporthe grisea and strategies for its management. Blast Disease of Cereal Crops: Evolution and Adaptation in Context of Climate Change. Springer; 2021. p. 151-172

[35] Singh Y, Kumar J. Study of genomic fingerprints profile of Magnaporthe grisea from finger millet (Eleusine Coracona) by random amplified polymorphic DNA-polymerase chain reaction (RAPD-PCR). African Journal of Biotechnology. 2010;**9**(46):7798-7804

[36] Panda D, Kundu SK, Ghosh A, Prakash NB, Patra DD. Agricultural nitrogen use and its environmental implications. In: Abrol YP, Raghuram N, Sachdev MS, editors. Nitrogen use efficiency in rice ecosystems. New Delhi: I.K. International Publishing House Pvt. Ltd; 2007. pp. 99-120

[37] Rao AN. Estimates of losses in finger millet (Eleusine coracana) due to blast disease (Pyricularia grisea). Mysore Journal of Agricultural Sciences. 1990;**24**(1):57-60

[38] Ramappa HK, Ravishankar CR, Prakash P, Improvement AI. Estimation of yield loss and management of blast disease in finger millet (ragi). Journal of Mycology and Plant Pathology (India). 2002;**32**(3):409

[39] Nayaka CS, Srivastava RK, Udayashankar AC, Lavanya SN, Prakash G, Bishnoi HR. Magnaporthe blast of pearl millet in India - present status and future prospects. All India Coordinated Research Project on Pearl Millet (Indian Council of Agricultural Research). 2017;**342**(304):1-51

[40] Lule D, de Villiers S, Fetene M, Bogale T, Alemu T, Geremew G, et al. Pathogenicity and yield loss assessment caused by Magnaporthe oryzae isolates in cultivated and wild relatives of finger millet (Eleusine coracana). Indian. Journal of Agricultural Research. 2014;**48**(4):258-268

[41] Sekar J, Raju K, Duraisamy P, Ramalingam VP. Potential of finger millet indigenous rhizobacterium pseudomonas sp. MSSRFD41 in blast disease management—Growth promotion and compatibility with the resident rhizomicrobiome. Frontiers in microbiology. 2018;**9**:1029

[42] Prajapati VP, Sabalpara AN, Pawar DM. Assessment of yield loss due to finger millet blast caused by Pyricularia grisea (Cooke) Sacc. Trends in Biosciences. 2013;**6**(3):876-788

[43] Anonymous. 2021. Annual Progress Report of Annual Progress Report of Indian Council of Agricultural research (ICAR)-All India Co-Ordinated Research Project (AICRP) on Small Millets. https://www.millets.res.in/

[44] Karthikeyan V, Gnanamanickam SS. Biological control of Setaria blast (Magnaporthe grisea) with bacterial strains. Crop Protection. 2008;**27**(2):263-267

[45] Das IK, Krishnan S, Tonapi VA. Current Status and Management of Foxtail Millet [Setaria italica (L.) Beauv.] Blast Disease. InBlast Disease

of Cereal Crops. Cham: Springer; 2021. pp. 201-209

[46] LongyaA,TalumphaiS,JantasuriyaratC. Morphological characterization and genetic diversity of rice blast fungus, Pyricularia oryzae, from Thailand using ISSR and SRAP markers. Journal of Fungi. 2020;**6**(1):38

[47] Lukose CM, Kaodvani D, Pethani KV. Factors inducing growth and sporulation in Pyricularia penniseti of pearl millet. Indian Phytopathology. 2012;**54**(4):484-485

[48] Konda S. Investigations on Leaf Blast of Foxtail Millet [*Setaria italica* (L.) Beauv.] [thesis]. Bengaluru: University of Agricultural Sciences; 2015

[49] Nabi A, Shah MD, Padder BA, Rasool F, Ahanger FA, Ahmad M. Molecular characterization and phylogenetic analysis of Magnaporthae oryzae BC couch causing blast disease of Rice (Oryza sativa) in Kashmir. SKUAST Journal of Research. 2015;**17**(1):15-22

[50] Sonah H, Deshmukh RK, Parida SK, Chand S, Kotasthane A. Morphological and genetic variation among different isolates of Magnaporthe grisea collected from Chhattisgarh. Indian Phytopathology. 2009;**62**(4):469

[51] Adhikari S, Joshi SM, Athoni BK, Patil PV, Jogaiah S. Elucidation of genetic relatedness of Magnaporthe grisea, an incitent of pearl millet blast disease by molecular markers associated with virulence of host differential cultivars. Microbial Pathogenesis. 2020;**149**:104533

[52] Kato H, Yamamoto M, Yamaguchi-Ozaki T, Kadouchi H, Iwamoto Y, Nakayashiki H, et al. Pathogenicity, mating ability and DNA restriction fragment length polymorphisms of Pyricularia populations isolated from Gramineae, Bambusideae and Zingiberaceae plants. Journal of General Plant Pathology. 2000;**66**(1):30-47

[53] Babu TK, Sharma R, Upadhyaya HD, Reddy PN, Deshpande SP, Senthilvel S, et al. Evaluation of genetic diversity in Magnaporthe grisea populations adapted to finger millet using simple sequence repeats (SSRs) markers. Physiological and Molecular Plant Pathology. 2013;**84**:10-18

[54] Hebert TT. The perfect stage of Pyricularia grisea. Phytopathology. 1971;**61**(1):83-87

[55] Palanna KB, Vinay KHD, Rajashekara H, Jeevan B, Raveendra HR, Patro SK, et al. Assessment of mating type alleles and fertility of *Magnaporthe* population adapted to millets in India. In: 8th International Conference (Hybrid mode) on Plant Pathology: Retrospect and Prospects. 2022. p. 27-49

[56] ICAR - All India Coordinated Research Project on Small Millets, Bengaluru. Available from: www.aicrpsm.res.in

[57] Kurahashi Y. Melanin biosynthesis inhibitors (MBIs) for control of rice blast. Pesticide Outlook. 2001;**12**(1):32-35

[58] Iwata M. Probenazole-a plant defence activator. Pesticide Outlook. 2001;**12**(1):28-31

[59] Uesugi Y. Fungal choline biosynthesis-a target for controlling rice blast. Pesticide Outlook. 2001;**12**(1):26-27

[60] Kato H. Rice Blast Control. Japan: Pesticide Outlook. Japan crop protection association; 2001. pp. 23-25

[61] https://www.fao.org/globalperspectivesstudies/foodagricultureprojections-to-2050/en/

[62] Salimath SS, Oliveira AC, Bennetzen JL, Godwin ID. Assessment of genome origins and genetic diversity in the genus Eleusine with DNA markers. Genome. 1995;**38**(4):757-763

[63] Ashkani S, Rafii MY, Rusli I, Sariah M, Abdullah SN, Abdul Rahim H, et al. SSRs for marker-assisted selection for blast resistance in rice (Oryza sativa L.). Plant Molecular Biology Reporter. 2012;**30**(1):79-86

[64] Saleem S, Mushtaq NU, Shah WH, Rasool A, Hakeem KR, Rehman RU. Morpho-physiological, biochemical and molecular adaptation of millets to abiotic stresses: A review. Phyton. 2021;**90**(5):1363

[65] Kawasaki S, editor. Rice Blast: Interaction with Rice and Control. Proceedings of the 3rd International Rice Blast Conference. Tsukuba Science City, Ibaraki, Japan: Kluwer Academic Publishers; 2004. pp. 302

[66] Mousa WK, Raizada MN. Natural disease control in cereal grains. Elsevier; 2016

[67] Ceasar A. Genome-editing in millets: Current knowledge and future perspectives. Molecular Biology Reports. 2022;**49**(1):773-781

[68] Hittalmani S, Mahesh HB, Shirke MD, Biradar H, Uday G, Aruna YR, et al. Genome and transcriptome sequence of finger millet (Eleusine coracana (L.) Gaertn.) provides insights into drought tolerance and nutraceutical properties. BMC Genomics. 2017;**18**(1):1-6

[69] Varshney RK, Shi C, Thudi M, Mariac C, Wallace J, Qi P, et al. Pearl millet genome sequence provides a resource to improve agronomic traits in arid environments. Nature Biotechnology. 2017;**35**(10):969-976

[70] Zou C, Li L, Miki D, Li D, Tang Q, Xiao L, et al. The genome of broomcorn millet. Nature Communications. 2019;**10**(1):1-11

[71] Sood S, Joshi DC, Chandra AK, Kumar A. Phenomics and genomics of finger millet: Current status and future prospects. Planta. 2019;**250**(3):731-751

[72] Shivhare R, Lata C. Exploration of genetic and genomic resources for abiotic and biotic stress tolerance in pearl millet. Frontiers in Plant Science. 2017;7:2069